2

ILLINOIS CENTRAL COLLEGE

A12901 809274

P9-DFE-184

COSTING THE EARTH?
Perspectives on Sustainable D

The United Nations are estimating that the global population will grow to 9 billion by 2050. Up to the year 2030, the gross domestic product of the world will increase by 130% – especially in the newly industrialized countries like China and India which are already displaying enormously rapid economic growth. The consequence: 50% more resources will be extracted from nature. There will be a similar increase in the deposit of harmful substances and garbage in nature, and the increased emission of greenhouse gases. Global warming of about 2 °C can no longer be avoided. This volume shows what options we have to deal with this, potentially the biggest challenge ever confronted by mankind, and how our economies have to react.

Bernd Meyer is Professor for Macroeconomics at the University of Osnabrück and Head of Science at the Society for the Study of Economic Structures. He was the head of the committee "Evolutionary Economics" at the Society of Economics and Social Sciences, and of the scientific advisory board on economic-environmental accounting for the German Ministry of the Environment.

Our addresses on the Internet:
www.the-sustainability-project.com
www.forum-fuer-verantwortung.de
[English version available]

COSTING THE EARTH?
Perspectives on Sustainable Development

BERND MEYER

Translated by Baker & Harrison
Klaus Wiegandt, General Editor

HAUS PUBLISHING

First published in Great Britain in 2009 by
Haus Publishing Ltd
70 Cadogan Place
London SW1X 9AH
www.hauspublishing.com

Originally published as: *Wie Muss die Wirtschaft Umgebaut Werden?*, by Bernd Meyer

Copyright © 2007 Fischer Taschenbuch Verlag in der S. Fischer Verlag GmbH, Frankfurt am Main

English translation copyright © Baker & Harrison 2008

The moral right of the author has been asserted

A CIP catalogue record for this book
is available from the British Library

ISBN 978-1-906598-12-9

Typeset in Sabon by MacGuru Ltd
Printed in Dubai by Oriental Press

Mixed Sources
Product group from well-managed
forests and other controlled sources
www.fsc.org Cert no. CU-COC-809367
© 1996 Forest Stewardship Council

Haus Publishing believes in the importance of a sustainable future for our planet. This book is printed on paper produced in accordance with the standards of sustainability set out and monitored by the FSC. The printer holds chain of custody.

Contents

Editor's Foreword

Sustainability Project

Sales of the German-language edition of this series have exceeded all expectations. The positive media response has been encouraging, too. Both of these positive responses demonstrate that the series addresses the right topics in a language that is easily understood by the general reader. The combination of thematic breadth and scientifically astute, yet generally accessible writing, is particularly important as I believe it to be a vital prerequisite for smoothing the way to a sustainable society by turning knowledge into action. After all, I am not a scientist myself; my background is in business.

A few months ago, shortly after the first volumes had been published, we received suggestions from neighboring countries in Europe recommending that an English-language edition would reach a far larger readership. Books dealing with global challenges, they said, require global action brought about by informed debate amongst as large an audience as possible. When delegates from India, China, and Pakistan voiced similar concerns at an international conference my mind was made up. Dedicated individuals such as Lester R. Brown and Jonathan Porritt deserve credit for bringing the concept of sustainability to the attention of the general public, I am convinced that this series can give the discourse about sustainability something new.

Two years have passed since I wrote the foreword to the initial German edition. During this time, unsustainable developments on our planet have come to our attention in ever more dramatic ways. The price of oil has nearly tripled; the value of industrial metals has risen exponentially and, quite unexpectedly, the costs of staple foods such as corn, rice, and wheat have reached all-time highs. Around the globe, people are increasingly concerned that the pressure caused by these drastic price increases will lead to serious destabilization in China, India, Indonesia, Vietnam, and Malaysia, the world's key developing regions.

The frequency and intensity of natural disasters brought on by global warming has continued to increase. Many regions of our Earth are experiencing prolonged droughts, with subsequent shortages of drinking water and the destruction of entire harvests. In other parts of the world, typhoons and hurricanes are causing massive flooding and inflicting immeasurable suffering.

The turbulence in the world's financial markets, triggered by the US sub-prime mortgage crisis, has only added to these woes. It has affected every country and made clear just how unscrupulous and sometimes irresponsible speculation has become in today's financial world. The expectation of exorbitant short-term rates of return on capital investments led to complex and obscure financial engineering. Coupled with a reckless willingness to take risks everyone involved seemingly lost track of the situation. How else can blue chip companies incur multi-billion dollar losses? If central banks had not come to the rescue with dramatic steps to back up their currencies, the world's economy would have collapsed. It was only in these circumstances that the use of public monies could be justified. It is therefore imperative to prevent a repeat of speculation with short-term capital on such a gigantic scale.

Taken together, these developments have at least significantly improved the readiness for a debate on sustainability. Many more are now aware that our wasteful use of natural resources and energy have serious consequences, and not only for future generations.

Two years ago, who would have dared to hope that WalMart, the world's largest retailer, would initiate a dialog about sustainability with its customers and promise to put the results into practice? Who would have considered it possible that CNN would start a series "Going Green"? Every day, more and more businesses worldwide announce that they are putting the topic of sustainability at the core of their strategic considerations. Let us use this momentum to try and make sure that these positive developments are not a flash in the pan, but a solid part of our necessary discourse within civic society.

However, we cannot achieve sustainable development through a multitude of individual adjustments. We are facing the challenge of critical fundamental questioning of our lifestyle and consumption and patterns of production. We must grapple with the complexity of the entire earth system in a forward-looking and precautionary manner, and not focus solely on topics such as energy and climate change.

The authors of these twelve books examine the consequences of our destructive interference in the Earth ecosystem from different perspectives. They point out that we still have plenty of opportunities to shape a sustainable future. If we want to achieve this, however, it is imperative that we use the information we have as a basis for systematic action, guided by the principles of sustainable development. If the step from knowledge to action is not only to be taken, but also to succeed, we need to offer comprehensive education to all, with the foundation in early childhood. The central issues of the future must be anchored firmly in school curricula, and no university student should be permitted to graduate without having completed a general course on sustainable development. Everyday opportunities for action must be made clear to us all – young and old. Only then can we begin to think critically about our lifestyles and make positive changes in the direction of sustainability. We need to show the business community the way to sustainable development via a responsible attitude to consumption, and become active within our sphere of influence as opinion leaders.

For this reason, my foundation *Forum für Verantwortung*, the ASKO EUROPA-FOUNDATION, and the European Academy Otzenhausen have joined forces to produce educational materials on the future of the Earth to accompany the twelve books developed at the renowned Wuppertal Institute for Climate, Environment and Energy. We are setting up an extensive program of seminars, and the initial results are very promising. The success of our initiative "Encouraging Sustainability," which has now been awarded the status of an official project of the UN Decade "Education for Sustainable Development," confirms the public's great interest in, and demand for, well-founded information.

I would like to thank the authors for their additional effort to update all their information and put the contents of their original volumes in a more global context. My special thanks goes to the translators, who submitted themselves to a strict timetable, and to Annette Maas for coordinating the Sustainability Project. I am grateful for the expert editorial advice of Amy Irvine and the Haus Publishing editorial team for not losing track of the "3600-page-work."

Taking Action — Out of Insight and Responsibility

"We were on our way to becoming gods, supreme beings who could create a second world, using the natural world only as building blocks for our new creation."

This warning by the psychoanalyst and social philosopher Erich Fromm is to be found in *To Have or to Be?* (1976). It aptly expresses the dilemma in which we find ourselves as a result of our scientific-technical orientation.

The original intention of submitting to nature in order to make use of it ("knowledge is power") evolved into subjugating nature in order to exploit it. We have left the earlier successful path with its many advances and are now on the wrong track, a path of danger with incalculable risks. The greatest danger stems from the unshakable faith of the overwhelming majority of politicians and business leaders in unlimited economic growth which, together with limitless technological innovation, is supposed to provide solutions to all the challenges of the present and the future.

For decades now, scientists have been warning of this collision course with nature. As early as 1983, the United Nations founded the World Commission on Environment and Development which published the Brundtland Report in 1987. Under the title *Our Common Future*, it presented a concept that could save mankind from catastrophe and help to find the way back to a responsible way of life, the concept of long-term environmentally sustainable use of resources. "Sustainability," as used in the Brundtland Report, means "development that meets the needs of the present without compromising the ability of future generations to meet their own needs."

Despite many efforts, this guiding principle for ecologically, economically, and socially sustainable action has unfortunately not yet become the reality it can, indeed must, become. I believe the reason for this is that civil societies have not yet been sufficiently informed and mobilized.

Forum für Verantwortung

Against this background, and in the light of ever more warnings and scientific results, I decided to take on a societal responsibility with my foundation. I would like to contribute to the expansion of public discourse about sustainable development which is absolutely essential. It is my desire to provide a large number of people with facts and contextual knowledge on the subject of sustainability, and to show alternative options for future action.

After all, the principle of "sustainable development" alone is insufficient to change current patterns of living and economic practices. It does provide some orientation, but it has to be negotiated in concrete terms within society and then implemented in patterns of behavior. A democratic society seriously seeking to reorient itself towards future viability must rely on critical, creative individuals capable of both discussion and action. For this reason, life-long learning, from childhood to old age, is a necessary precondition for realizing sustainable development. The practical implementation of the ecological, economic, and social goals of a sustainability strategy in economic policy requires people able to reflect, innovate and recognize potentials for structural change and learn to use them in the best interests of society.

It is not enough for individuals to be merely "concerned." On the contrary, it is necessary to understand the scientific background and interconnections in order to have access to them and be able to develop them in discussions that lead in the right direction. Only in this way can the ability to make appropriate judgments emerge, and this is a prerequisite for responsible action.

The essential condition for this is presentation of both the facts and the theories within whose framework possible courses of action are visible in a manner that is both appropriate to the subject matter and comprehensible. Then, people will be able to use them to guide their personal behavior.

In order to move towards this goal, I asked renowned scientists to present in a generally understandable way the state of research and the possible options on twelve important topics in the area of sustainable development in the series "*Forum für Verantwortung.*" All those involved in this project are in agreement that there is no alternative to a united path of all societies towards sustainability:

- *Our Planet: How Much More Can Earth Take?* (Jill Jäger)
- *Energy: The World's Race for Resources in the 21st Century* (Hermann-Joseph Wagner)
- *Our Threatened Oceans* (Stefan Rahmstorf and Katherine Richardson)
- *Water Resources: Efficient, Sustainable and Equitable Use* (Wolfram Mauser)
- *The Earth: Natural Resources and Human Intervention* (Friedrich Schmidt-Bleek)
- *Overcrowded World? Global Population and International Migration* (Rainer Münz and Albert F. Reiterer)
- *Feeding the Planet: Environmental Protection through Sustainable Agriculture* (Klaus Hahlbrock)
- *Costing the Earth? Perspectives on Sustainable Development* (Bernd Meyer)

- *The New Plagues: Pandemics and Poverty in a Globalized World* (Stefan Kaufmann)
- *Climate Change: The Point of No Return* (Mojib Latif)
- *The Demise of Diversity: Loss and Extinction* (Josef H Reichholf)
- *Building a New World Order: Sustainable Policies for the Future* (Harald Müller)

The public debate

What gives me the courage to carry out this project and the optimism that I will reach civil societies in this way, and possibly provide an impetus for change?

For one thing, I have observed that, because of the number and severity of natural disasters in recent years, people have become more sensitive concerning questions of how we treat the Earth. For another, there are scarcely any books on the market that cover in language comprehensible to civil society the broad spectrum of comprehensive sustainable development in an integrated manner.

When I began to structure my ideas and the prerequisites for a public discourse on sustainability in 2004, I could not foresee that by the time the first books of the series were published, the general public would have come to perceive at least climate change and energy as topics of great concern. I believe this occurred especially as a result of the following events:

First, the United States witnessed the devastation of New Orleans in August 2005 by Hurricane Katrina, and the anarchy following in the wake of this disaster.

Second, in 2006, Al Gore began his information campaign on climate change and wastage of energy, culminating in his film *An*

Inconvenient Truth, which has made an impression on a wide audience of all age groups around the world.

Third, the 700-page Stern Report, commissioned by the British government, published in 2007 by the former Chief Economist of the World Bank Nicholas Stern in collaboration with other economists, was a wake-up call for politicians and business leaders alike. This report makes clear how extensive the damage to the global economy will be if we continue with "business as usual" and do not take vigorous steps to halt climate change. At the same time, the report demonstrates that we could finance countermeasures for just one-tenth of the cost of the probable damage, and could limit average global warming to 2° C – if we only took action.

Fourth, the most recent IPCC report, published in early 2007, was met by especially intense media interest, and therefore also received considerable public attention. It laid bare as never before how serious the situation is, and called for drastic action against climate change.

Last, but not least, the exceptional commitment of a number of billionaires such as Bill Gates, Warren Buffett, George Soros, and Richard Branson as well as Bill Clinton's work to "save the world" is impressing people around the globe and deserves mention here.

An important task for the authors of our twelve-volume series was to provide appropriate steps towards sustainable development in their particular subject area. In this context, we must always be aware that successful transition to this type of economic, ecological, and social development on our planet cannot succeed immediately, but will require many decades. Today, there are still no sure formulae for the most successful long-term path. A large number of scientists and even more innovative entrepreneurs and managers will have to use their creativity and dynamism to solve the great challenges. Nonetheless, even today, we can discern the first clear goals we must reach in order to avert a looming catastrophe. And billions of consumers around the world can use their daily purchasing decisions to help both ease and significantly accelerate the economy's transition to sustainable development – provided the political framework is there. In addition, from a global perspective, billions of citizens have the opportunity to mark out the political "guide rails" in a democratic way via their parliaments.

The most important insight currently shared by the scientific, political, and economic communities is that our resource-intensive Western model of prosperity (enjoyed today by one billion people) cannot be extended to another five billion or, by 2050, at least eight billion people. That would go far beyond the biophysical capacity of the planet. This realization is not in dispute. At issue, however, are the consequences we need to draw from it.

If we want to avoid serious conflicts between nations, the industrialized countries must reduce their consumption of resources by more than the developing and threshold countries increase theirs. In the future, all countries must achieve the same level of consumption. Only then will we be able to create the necessary ecological room for maneuver in order to ensure an appropriate level of prosperity for developing and threshold countries.

To avoid a dramatic loss of prosperity in the West during this long-term process of adaptation, the transition from high to low resource use, that is, to an ecological market economy, must be set in motion quickly.

On the other hand, the threshold and developing countries must commit themselves to getting their population growth under control within the foreseeable future. The twenty-year Programme of Action adopted by the United Nations International Conference on Population and Development in Cairo in 1994 must be implemented with stronger support from the industrialized nations.

If humankind does not succeed in drastically improving resource and energy efficiency and reducing population growth in a sustainable manner – we should remind ourselves of the United Nations forecast that population growth will come to a halt only at the end of this century, with a world population of eleven to twelve billion – then we run the real risk of developing eco-dictatorships. In the words of Ernst Ulrich von Weizsäcker: "States will be sorely tempted to ration limited resources, to micromanage economic activity, and in the interest of the environment to specify from above what citizens may or may not do. 'Quality-of-life' experts might define in an authoritarian way what kind of needs people are permitted to satisfy." (*Earth Politics*, 1989, in English translation: 1994).

It is time

It is time for us to take stock in a fundamental and critical way. We, the public, must decide what kind of future we want. Progress and quality of life is not dependent on year-by-year growth in per capita income alone, nor do we need inexorably growing amounts of goods to satisfy our needs. The short-term goals of our economy, such as maximizing profits and accumulating capital, are major obstacles to sustainable development. We should go back to a more decentralized economy and reduce world trade and the waste of energy associated with it in a targeted fashion. If resources and energy were to cost their "true" prices, the global process of rationalization and labor displacement will be reversed, because cost pressure will be shifted to the areas of materials and energy.

The path to sustainability requires enormous technological innovations. But not everything that is technologically possible has to be put into practice. We should not strive to place all areas of our lives under the dictates of the economic system. Making justice and fairness a reality for everyone is not only a moral and ethical imperative, but is also the most important means of securing world peace in the long term. For this reason, it is essential to place the political relationship between states and peoples on a new basis, a basis with which everyone can identify, not only the most powerful. Without common principles of global governance, sustainability cannot become a reality in any of the fields discussed in this series.

And finally, we must ask whether we humans have the right to reproduce to such an extent that we may reach a population of eleven to twelve billion by the end of this century, laying claim to every square centimeter of our Earth and restricting and destroying the habitats and way of life of all other species to an ever greater degree.

Our future is not predetermined. We ourselves shape it by our actions. We can continue as before, but if we do so, we will put ourselves in the biophysical straitjacket of nature, with possibly disastrous political implications, by the middle of this century. But we also have the opportunity to create a fairer and more viable future for ourselves and for future generations. This requires the commitment of everyone on our planet.

Klaus Wiegandt

Summer 2008

1 Introduction

An outline

A mid-ranged United Nations forecast variant envisages a world population of 9 billion by the year 2050. Population growth will take place in developing countries and the so-called emerging economies, where the population is expected to increase by approximately 50% by the year 2050. The emerging economies include developing countries, such as China and India, with very dynamic economic developments and annual gross domestic product (GDP) growth recorded between 6 to 10%.

By the year 2030 the world gross domestic product (GDP) will have grown by 130%. This means that despite an expected improvement in efficiency, the extraction of natural raw materials will increase by 50%.

The raw materials extracted from nature are processed into goods. Some of these goods take the shape of relatively durable buildings, machinery, roads and other facilities, while others will be passed on to the environment in the form of residual and harmful substances. Both the extraction of natural resources and more particularly the storage of residual and harmful substances in the natural environment causes lasting damage. In spite of all of the warnings scientists have issued, society has tolerated all of these facts with remarkable indifference to date.

It is only in view of the greenhouse effect resulting from

greenhouse gas emissions that the general public has started taking notice in the meantime. The accumulation of CO_2, methane and other gases in the earth's atmosphere act like the roof of a greenhouse: the Earth's temperature increases with the concentration of the so-called greenhouse gases in the atmosphere. A rise in temperature by 2 degrees Celsius is already unavoidable. The consequences are already tangible and will continue to intensify in the near future: mild winters and hot summers, more frequent and intense storms, a rise in sea levels, extinction of certain species and further unforeseeable reverberations. CO_2 is the most significant climate gas, which is produced when carbonaceous energy sources coal, gas and oil, are burnt, and then collects in the atmosphere. Unless we succeed in dramatically changing our behavior, by the year 2030 the expected economic growth will spell an increase in energy consumption and a further rise in CO_2 emissions of 40%. By the second half of this century the Earth's mean temperature will have risen not by 2 but 4 or 5 degrees. The repercussions are almost inestimable. To get an idea, one only has to consider that the difference between today's temperature and temperature during the ice ages is a mere 5 degrees.

What options do we have to meet what is probably the greatest of all challenges to mankind? One thing is certain: we need a global perspective. It is not enough to consider solutions only for Europe. We have to accept the realities of global development, population growth and economic growth in both developing countries and emerging market economies. Population growth in the third world has many cultural and socioeconomic aspects, which we will not be able to influence over the space of a few years. In addition, the United Nation's mid-range forecast variant cited above already assumes a decline in population growth. Economic growth is particularly welcomed by the developing world and emerging market economies, especially considering the drastic poverty still present today, and the prospects of economic growth lends hope for an improvement to the catastrophic social conditions. Development restrictions will certainly not be welcomed by these countries; after all, it has been mainly the industrialized countries responsible for polluting the environment to date.

If the developments described above are to be prevented, the concentration of climate gases must be stabilized by the second half of the century. This means that by then CO_2 emissions will have had to be globally reduced to 20% of 1990s levels, as this is the rate that correlates with the annual CO_2 assimilation from plant photosynthesis and other influences.

The only realistic option that we have at our disposal is a dramatic increase in the efficient use of raw materials. This means that we need a massive increase in the amount of goods produced per unit of raw material. Or to put it another way: the amount of raw materials consumed per unit of goods produced needs to be drastically reduced in order to help separate economic growth from resource consumption. On the one hand, this can be achieved by technical innovation and on the other hand by changing consumer habits. We do not have to consume less goods, we only need different goods that directly, and indirectly, contain less resources than have previously been used, and the commodities need to be manufactured by way of better, resource saving technologies. We will only be able to solve the problem with innovations that generate new consumer goods and production methods, as well as the necessary investments in buildings and machines. This is, however, similar to riding a tiger's back, because innovations and investment create economic growth, and yet they can only flourish in a dynamic economic environment.

But isn't this a contradiction? Doesn't growth mean more

consumption and more resource consumption? So far, the argument here has been that overall economic growth should not automatically mean more consumption of resources, because on the one hand we want to employ resource-saving technologies, and on the other we want to reduce the demand for the kind of consumer goods most responsible for high resource expenditure, and to create demand for more resource saving goods. This is what is meant by changes in consumer structures. The household's use of resources is sufficient, which means that they are restricting their resource consumption. As we see it this doesn't mean sufficiency *per se*. Occasionally the term 'sufficiency' is used closely in overall consumption, which is in my opinion problematic, because – as is yet to be shown – courses of action are restricted. The central proposition in this book is that politics has to adopt innovative strategies using two components. Politics needs to encourage companies to employ new resource saving production methods. In this context we speak of an efficiency strategy. In addition, consumers need to be persuaded to replace resource intensive goods. In this context, this is what we call a sufficiency strategy. Both components of the innovation strategy increase resource efficiency and decouple economic growth from resource consumption.

Thanks to its economic structure, Europe can be the driving force behind such a strategy. The CO_2 emission targets set during the German council presidency are an important step in this direction. Europe has to forge ahead with technological transformation in order to provide the world with improved resource saving technologies through international trade channels. Furthermore, hopes are pinned on Europe's example paving the way for worldwide target agreements and thereby ushering in the appropriate changes of behavior. During the 2007 G8 summit in Heiligendamm, and later in 2008 in Toyko government leaders from eight of the most economically important industrial countries set a target to cut global CO_2 emissions by at least a half by the year 2050. The heads of state and government gathered at Heiligendamm jointly agreed to integrate this target into the UN program. At the G8 summit in Toyko, the targets were confirmed. The main emerging market economy countries will also be included in this process. Should these words be put into action, this development will give hope that it might be possible to change course after all. A major development, compared to previous discussions, is America's participation, and the general opinion is that steps should be taken to encourage cooperation with the emerging market economies under the umbrella of the United Nations. Let us not forget that in 2006 talk of cutting climate gases by 50% by the year 2050 had only been discussed in the most optimistic scientific scenarios. Such a path is also economically beneficial for Europe, because our area is a leading global manufacturer of capital goods and holds a very strong position in the market of resource saving technologies.

This book provides a detailed account of innovative strategies geared to increasing resource efficiency, and shows the potentials as well as the risks. It will become clear that through an expedient combination of economic as well as regulatory policy instruments, more sustainable developments are possible. We reach the conclusion that there is no alternative path for the future, and we are also convinced that risks can be avoided by the appropriate accompanying measures.

Chapter contents

Chapter 2 discusses the question we have just addressed in detail: what awaits the world if we are unable to succeed in making dramatic changes to our behavior? Detailed explanations are given on population growth, economic growth and global resource consumption.

In Chapter 3 we ask why economic development is destroying the environment. The answer to this question offers different perspectives to the solution of this problem. We are able to use nature without incurring costs. As consumers or as producers we can, for example, emit pollutants into the Earth's atmosphere without facing costs as a result. This is the reason why we overuse nature and cause damage. It is also said that there is a divergence between the private price of using nature, and the social costs arising as a result of damage to the environment. In a complex modern political economy this can have fatal consequences, because decisions are made based on the wrong or faulty price of goods – faulty because the costs on nature have been neglected. The more we directly and indirectly use the primary products found in nature for our consumer goods, the more inaccurate or faulty the product's price is. How expensive would the computer that wrote these lines be if its price contained the cost of all direct and indirect damage impacting the environment during its production? Copper ore and other raw materials were extracted from the Earth, using methods detrimental to the environment. Air pollutants were generated when the ore was smelted, and the transportation of both the raw materials and the finished product also caused CO_2 emissions.

Market economy instruments are geared to charging those responsible for the costs to society resulting from damage to the environment. In the process, incentives are given to avoid these costs and reduce environmental damage. The so-called regulatory law operates using the instruments of prohibition and regulation. We will investigate the efficiency of both groups, and conclude that it all depends which specific problem needs solving. We are in the position where market economy instruments should admittedly be given priority, which will only be viable in connection with regulatory law, however.

After having established the basic position concerning the causes of environmental problems and the debate over possible general solutions, in Chapter 4 we will look into which objectives need to be pursued in the field of environmental policy. The debate on objectives will undoubtedly be dominated by the normative concept of sustainable development. In 1987, the World Commission on Environment and Development, chaired by the former Prime Minister of Norway Gro Harlem Brundtland, presented a report of its results. In this report a development is considered to be sustainable if it meets present day needs, without putting the needs of future generations at risk. The term sustainability has been a focus of environmental debate for the last twenty years, and has received various interpretations and comments. More than anything, efforts have been taken for a clear definition in order to set clear terms of reference for policy making.

Sustainability is an anthropocentric concept: mankind and its needs are central. Sustainable development has an ecological, an economical and a social dimension. Both the ecological and economical dimensions of sustainability are concerned with passing on a certain capital stock to the next generation, both natural and economic. Nature's capital stock consists of air, seas, rivers, earth, ecosystems, biodiversity, mineral resources – an inexhaustible list. The economic capital stock includes, more than anything, buildings and machinery, but also the human capital, including knowledge and experience. Naturally the quality of natural capital and economic capital should remain preferably unchanged, whereby 'capital' is initially meant in the physical sense of the term without any monetary values.

It is slightly harder to apply the term 'capital' to the social aspects of sustainability, because social sustainability only denotes a certain distribution of goods. However, the term capital has since established itself in this context. Social capital is the term used for the results achieved by specific institutions responsible for maintaining social balance. Among these are social laws within the frames of regulatory laws, redistribution via the tax system and social security, and also negotiation practices between representatives of various social groups. The presence of such institutions ensures a certain level of social balance, and because of this it is considered to be a community's social capital.

Now one could – admittedly somewhat abstractly – envisage a sustainable development where one generation passes its social, natural and economic capital on to the next generation in the same condition as when they received it, the same quality and quantity. This image immediately raises the question of whether the size of each capital should be examined individually, or whether it is more important to examine the total size of all three combined for evaluation. This is the point where two concepts of sustainability come into play – 'weak' sustainability and 'strong' sustainability.

The concept of weak sustainability is concerned with the preservation of total capital stock. This means that a loss in natural capital could be substituted with an increase in economic capital. This concept should be rejected because in some cases it could pose a threat to mankind's livelihood. Strong sustainability does not tolerate any form of substitution between the three dimensions. In its most extreme form it does not even allow substitution within a particular dimension. This is obviously going a little too far, because then the extraction of non-renewable resources such as mineral ores, fossil energy sources would be totally prohibited.

This would make the economic process as we know it today obsolete. Consequently, a mild form of strong sustainability has been chosen. Any form of substitution within natural capital is permissible. Areas could be sealed with asphalt, providing that other areas are converted into new landscape conservation areas or nature reserves. At the same time there has to be a guarantee that the main eco-systems are able to work. This variation would appear to be ethical as well as viable. Germany's environmental policy practices, and those pursued in most other European countries, are based on this interpretation of sustainability.

The reason we are engaging in a debate that would initially appear to be very theoretical is that it would be impossible to formulate long-term development policies without their orientation to objectives and targets. Environmental policy's leaning towards sustainability has very important practical consequences. The traditional policy of maintenance, which has been concerned with removing consumer and production waste from the environment, can only be described as being completely insufficient. The primary concern had been the attachment of filters to prevent emissions, and had nothing to do with changing our behavior. Yet this is exactly what a sustainable strategy requires. How else will it be possible to sustain resources and simultaneously provide enough employment, which, in connection with constant technical innovation and economic development implies an increase in resource consumption?

The necessary behavioral changes can be narrowed down to one central factor: an increase in resource productivity. This means the total of goods produced per resource unit. This is how economic growth can be decoupled from resource consumption. The more effectively, for example, mineral ore or oil are used, the fewer resource units are needed, and the greater the benefits in goods production. Basically there are two strategies available to reach this goal: the so-called sufficiency strategy aims to change consumer habits. We need to try to replace goods which involve a high consumption of resources in connection with existing production technologies. For example, less individual transportation in private vehicles and consequently more public transport services such as trains. The efficiency strategy favors technical innovation. Staying with our example: fuel consumption in cars, and electricity consumption in trains need to be reduced. Both strategies need to be applied.

In Chapter 5 we will be questioning the potentials of efficiency and sufficiency strategies from a macroeconomic point of view. It will become clear that resource consumption is strongly concentrated around certain consumer goods and also specific technologies. This is good news, because it implies that relatively small changes in consumer habits and in the production methods employed are capable of achieving relatively major impact on resource utilization. Further good news is that a number of resource saving technologies, which can be used directly or indirectly in the production of most goods – this is why they are called cross-section technologies – are expected to be highly successful in the near future. These technologies include information and communication technologies, nanotechnology, biotechnology and renewable energies.

How can the desired technical innovations and changes in consumer behavior be achieved, and also how can an already visible increase in raw material productivity be accelerated? These questions will be answered in Chapter 6 with the debate of concrete measures. We begin with economic instruments and discuss further developments in Europe's already implemented emissions trading with CO_2 certificates, the taxation of companies and households with eco-taxes, as well as the promotion of innovations with subventions, and also by employing consulting outfits and information agencies. The certification of consumer goods, durable goods and buildings based on their ecological properties seems to be important in helping provide businesses and households with the information necessary to react appropriately to economic instruments. An alternative could include a specification of technical norms for each of the best technologies available for vehicles, equipment and buildings, and which would have to be met by manufacturers within a specific time frame.

The instruments mentioned so far endeavor to provide incentives (economic instruments) and constraints (regulatory law) in order to reorganize the current economy to a sustainable economy making more prudent use of resources. The most ethical path is, of course, to change behavior by persuasion. It is a question of an intrinsic motivation to improve resource productivity, arising from a self-initiated awareness of its necessity. In this context the chapter discusses educating the population on environmental economical interactions, and whether more sustainable economic management styles are supportable.

We have already argued for a dynamic reorganization of the economy, which envisions Europe, as the driving force behind change. We have given detailed separate descriptions of which measures should be used to encourage companies' willingness for innovation. In the process it has become clear that this concept can only be realized within the framework of international free competition. As we have already seen in the first chapter, technological transition always requires changes in societal structures too, and can often cause uncertainties. The German population for example is already concerned because of the risks of job loss increase with international competitiveness. On the other hand, a successful German export trade and industry has always offered new opportunities for well-qualified and flexible employees. This problem here is that while our strategy improves the opportunities for the more successful individuals, the risks for the less qualified increase.

What are the changes we will have to adapt to as far as job qualification requirements are concerned over the next few decades? What will the effects of the demographic changes over the next 20 years look like, and what role will economic changes play within this context? We will follow these questions in Chapter 7 and arrive at the conclusion that without a campaign to improve education, in 20 years there will be a large deficit of highly qualified workers and a surplus of unskilled workers on the employment market. The latter will require securing the wages of the unskilled, which cannot be achieved with minimum wage regulations, but through state funded transfers to top up available income of low earners. We expect that an innovative strategy containing an education campaign and secure income will enable a more sustainable development from an ecological, economical and social aspect.

Chapter 8 discusses the results of an EU Commission research project to investigate what effects an innovative strategy would have on Europe's economy and environment. The results of the simulation calculations using a global model, illustrate the repercussions of economical development on the environment, and show that the necessary economical reorganization of Europe will need considerable effort. It becomes clear that success in Europe alone is nowhere near enough to solve the global problem.

Chapter 9 deals with the necessity of an international coordination in environmental policy. At present, climate protection is the central topic in international environmental policy debates. The agreements reached on climate protection in 1997 – otherwise known as the Kyoto Protocol – regulate tolerable greenhouse gas emissions within industrial countries. The United States' non-participation had been a decisive weakness in the

agreement, which expires in 2012. The United States' participation in the Heiligendamm and Toyako declarations has given a
strong impetus in efforts to come up with a successor to the Kyoto
Protocol. Now it is all down to some kind of inclusion of the
emerging market economies of China and India. The difficulties
in negotiations are due to the fact that neither of the countries
have been really that responsible for polluting the atmosphere
up to now, and present emission rates per head are very low. On
the other hand, it is clear that due to these countries' enormous
economic growth a decoupling of economic developments and
CO_2 emissions is necessary. These countries also have to increase
resource productivity, and with the help of western industrial
countries it might be possible to set target agreements.

2 Where is the World Floating?

Economic growth and competition between industrialized countries and emerging developing countries

Developments in the global economy have been increasingly influenced by the phenomenon of globalization over the last twenty-five years. Individual national economies are firmly integrated into global economic development processes, from which they are no longer able to distance themselves. These developments have enabled enormous progress in information technologies, permitting unrestricted (both spatial and quantitative) international transactions today. At the same time the world economy has been liberalized: China has, at least partially, opened up to market based structures, and in Eastern Europe former socialist countries have transformed into market economies. The world capital market has opened up completely new dimensions to the international division of labor. Investors in industrial countries are seizing the opportunity to build production sites in China, India, and Southeast Asia or in Eastern Europe, where wages are considerably lower than those of the industrial countries. A big proportion of these products are exported back into industrial countries as a result of international commodity trading, either to be used as consumer goods, or as primary products to be further developed. Location plays a decisive role, and every time a corporation is about to expand, today's manufacturers are faced with the question of relocation.

The coverage of such matters in the media is often accompanied by comments expressing concerns over job losses. Yet it is often overlooked that a more widespread distribution of international labor also offers advantages to industrialized countries. Aside from the costs of manpower, there are other decisive factors to be taken into consideration: workers qualifications, local traffic infrastructure and research facilities, the proximity to the customers of the product, and legal security on the production site. In the particular case of highly technical products, it has often proven to be of an advantage when there is a concentrated group of users of specific technologies. The term used here is industrial "clustering," and it can offer long-term advantages with regard to the realization of technical developments. Economic growth in the developing countries and the newly industrialized countries – as the more successful developing countries are now called – also offers an increased demand potential for the industrial countries. The consequent development processes in the changes in the global division of labor leads to increased competition. For everyone involved in this process – whether they are producers, investors or employees – this spells increased risks, and opportunities. This is nothing new for employers, but the higher risks of job losses, and the consequent pressure on wage rates is perceived to be a threat to the living standards of employees.

In fact not every country or region profits from globalization. It all depends on whether the strengths can be exploited in this competitive process, and if the weaknesses can be minimized. This calls for permanent structural change, constantly creating new industries, products, and occupational skills to replace the old. Those who accept this are likely to flourish, but at the same

| | 1980– | 1990– | 2004– | 2015– | 2004– |
	1990	2004	2015	2030	2030
OECD	3.0	2.5	2.6	1.9	2.2
North America	3.1	3.0	2.9	2.0	2.4
USA	3.2	3.0	2.9	1.9	2.3
Europe	2.4	2.2	2.3	1.8	2.0
Pacific	4.2	2.2	2.3	1.6	1.9
Japan	3.9	1.3	1.7	1.3	1.4
Transf. countries	−0.5	−0.8	4.4	2.9	3.6
Russia		−0.9	4.2	2.9	3.4
Developing countries	3.9	5.7	5.8	3.9	4.7
in Asia	6.6	7.3	6.4	4.1	5.1
China	9.1	10.1	7.3	4.3	5.5
India	6.0	5.7	6.4	4.2	5.1
Middle East	−0.4	3.9	5.0	3.2	4.0
Africa	2.1	2.8	4.4	3.6	3.9
Latin America	1.3	2.8	3.5	2.9	3.2
Brazil	1.5	2.6	3.3	2.8	3.0
World	2.9	3.4	4.0	2.9	3.4
EU	2.4	2.1	2.2	1.8	2.0

Table 1 Global real gross domestic product growth rates
Average annual growth rates in percent
Source: International Energy Agency: World Energy Outlook 2006

time will have to accept fast paced economical and social structural changes.

The International Energy Agency (2006) (IEA) has calculated
the growth of the gross world product over the past twenty-five
years, and estimated future developments up to the year 2030.
This data is illustrated in Table 1. The global world product is
the total value of products produced worldwide over a specific
period. The various currencies were converted using purchasing power parities. Investigations were made to determine how
many units of various countries' and regions' uniform baskets

could be bought with the respective gross domestic product.

China, India and Southeast Asia in particular have profited from globalization over the past twenty-five years. From 1980 to 2004 China was able to increase its annual GDP by almost 10%, whereas industrial countries belonging to the OECD (Organization for Economic Cooperation and Development) achieved noticeably weaker economic growth with rates from between 2.5% and 3%. India and Southeast Asian countries not belonging to the OECD show growth rates between 6% and 7%, which are also noticeably above the industrial countries' growth rates. The total annual growth rate of the developing countries stands between 4% and 6%, whereby Africa and Latin America's developments are below average. The IEA expects a slight slowing down of Asia's growth, and increases for Asia and Latin America, which would mean an average annual growth rate of 4.7%. The industrial countries (OECD) are expected to grow only by an average 2.2% a year. Despite these wide differences in annual growth rates, in 2030 the per capita income of the OECD countries will still be four times greater than the rest of the world, because on the one hand the differences in today's incomes are dramatic, and on the other hand there will be an increase in the population of the developing and newly industrialized countries. Nevertheless, at least the difference between the per capita income of the industrial countries and the developing countries will decrease. By 2030, the annual gross world product will continue growing to a total 3.4%. While predictions of future stable economic development are comforting on the one hand, they are alarming in view of the environmental problems.

Globalization has had a strong influence on Europe and especially on Germany's economic development, because the country traditionally has strong connections with the global economy.

Table 2 illustrates this with export and import shares showing

	1995	2007
Export quota	24.0	46.7
Import quota	23.5	39.7

Table 2 Germany's export and import quotas
Source: Federal Statistical Office

the relation of both exports and imports to the GDP from 1995 up to 2007. Within just twelve years the export ratio has risen from 24% to 46.7%. Apart from the world economic dynamic just mentioned, this phenomenon can be explained by the important role of the progressive integration of the Euro zone and Europe's extension into Eastern Europe. Germany is a world champion of exports, exporting more goods than the United States, whose GDP exceed Germany's by around five times. On the other hand, the German import ratio has also increased from 23.5% to 39.7%. But a gap has appeared between the two variables, the so-called net export, which today amounts to 7% of the GDP, or 170.9 billion euros. Even more remarkable is the fact that approximately 60% of German exports consist of capital goods. If we were to add the exports of chemical products, this would then make up a 72% share of the total exports in this group of commodities. So there are then only just a few industries generating Germany's exports, such as mechanical engineering and machine building, automotive industry, electrical engineering, control engineering or chemistry. Measured against the domestic demand for producer goods and chemical products, these industries are extremely oversized. Germany produces these capital goods and chemical goods for the world. This explains the dramatic increase of German exports resulting from the extension of the production capacities of Eastern Europe and newly industrialized countries. Table 3 shows Germany's

Engineering	15.6%
Automotive Industry	19.1%
Electrical Engineering	10.6%
Chemicals	11.0%

Table 3 Germany's percentage of global imports of selected commodity
groups in 2002
Source: OECD

share of international imports in 2004 according to commodity groups. Every fifth vehicle and every sixth machine imported internationally originates from Germany.

Of course, high dependency on exports also has its risks, especially since this includes a heavy specialization in Germany of the production of capital and chemical goods. Then again, it also offers opportunities, because the production of these complex technical products requires specific knowledge. Furthermore, these products contain a considerable amount of primary products, which creates added value and employment to many other industries.

Continuous population growth

The United Nations are continuously working out world population forecasts, divided by country, and updated every two years. The following details are based on the World Population Prospects (2005). The way a country's population develops all depends on natural population movement and migration. Natural population movement is governed by mortality and fertility. Mortality equals life expectancy, according to age and gender, fertility is defined as the average number of children born per woman. A fertility rate of 2.1 means that population figures stay constant.

If, however, the rate is higher, population figures rise, and a fertility rate below 2.1 means a shrinking population.

Both rates will change in the future. Improvements in medical care and a general improvement of living conditions as a result of an increase in wealth will mean lower mortality, which leads to higher life expectancy. On the other side, epidemic illnesses such as AIDS push the mortality rate higher. Both of these factors have been taken into consideration in the United Nation's calculations. The fertility rate is the main factor in long-term natural population development, and is decisively influenced by prosperity. The states of poorer economies are often less able to provide social security. It is the family that is responsible for covering the risks that illnesses pose to income, and also for supporting the elderly. Consequently, having more children seems to be more attractive. Then again, with an increase in job opportunities and a raise in a nation's income, there is also an increased desire to profit from these developments, leading to less time being made available in order to raise children because of an increased amount of time spent earning a living. These correlations have been observed in both developing and industrial countries, albeit at different levels.

So in long term forecasts for 2050, due to expected economic growth, it can be assumed that there will be a decrease in the fertility rate. However, the exact scale of this decrease is still unclear. Consequently, the United Nations have prepared four variations in their projections, all with different fertility levels.

Figure 1 shows the historical development of the fertility rate since 1950, and mid-range estimates for the industrial countries, developing countries, the newly industrialized countries, the poorest countries, and for the world in total. The historical development impressively underlines the thesis of correlations between economic development and the fertility rate in

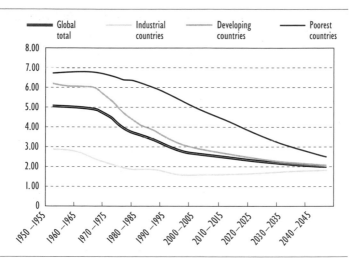

Figure I   The development of the fertility rate for the world, and
divided into groups of countries, from the mid-range
projections of the United Nations

two ways: first of all we can see that the highest fertility rate is
in the poorest countries, followed by the developing and newly
industrialized countries, and the lowest rates are in the industrial countries. Secondly, in all groups the fertility rate drops as
time progresses. This is especially noticeable in the developing
and newly industrialized countries where, starting in the sev-
enties, there has been a dramatic decrease in the fertility rate.
The decline in fertility spreads into the developing countries,
the newly industrialized countries, and the poorest countries.
In contrast, the UN projections of fertility in industrial countries predict a gradual increase to just fewer than 2.0. This could
be interpreted as society's unwillingness to accept a shrinking
population, eventually creating socio-economic structures to

Areas	Population in millions			Population in 2050 in millions			
	1950	1975	2005	low	medium	high	constant
World	2519	4074	6465	7680	9076	10646	11658
Industrialized countries	813	1047	1211	1057	1236	1440	1195
Developing countries	1707	3027	5253	6622	7840	9206	10463
Poorest countries	201	356	759	1497	1735	1994	2744
Other developing countries	1506	2671	4494	5126	6104	7213	7719
Africa	224	416	906	1666	1937	2228	3100
Asia	1396	2395	3905	4388	5217	6161	6487
Europe	547	676	728	557	653	764	606
Latin America and the Caribbean	167	322	561	653	783	930	957
North America	172	243	331	375	438	509	454
Oceania	13	21	33	41	48	55	55

Table 4 World population developments according to groups of
countries and various fertility rate assumptions
Source: Population Division of the Department of Economic and Social Affairs
of the United Nations Secretariat (2005) World Population Prospects: The 2004
Revision. Highlights. New York: United Nations

help stabilize population figures. This is reflected in the present
debate in Germany on demographic changes, and the necessity
for child-friendly policies.

Table 4 illustrates a summary of total world population and
regional developments, including all four variations of fertility
developments.

The mid-range variant estimates that current fertility, aver-
aging 2.6 children per woman (in all countries), will shrink to

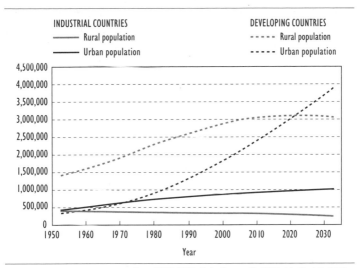

INDUSTRIAL COUNTRIES
— Rural population
— Urban population

DEVELOPING COUNTRIES
- - - - - Rural population
- - - - Urban population

Source: UN World Urbanization Prospects: The 2005 Revision Population Database

Figure 2   The development of rural and urban populations of the industrial countries and developing countries, in thousands of people

just above two by the year 2050. In the high variant only a slight decrease in fertility, to 2.5, is estimated, and in the low variant a rate of 1.5. Obviously the fertility rates in each country are different. If we assume that fertility remains constant, the highest population growth rate of 80% would equal an increase of today's 6.5 billion to 11.7 billion in 2050. Yet even with a dramatic decline in the world's average fertility of 1.5 in 2050 – a figure that lays just above Germany's current fertility of 1.4 – there would still be a further increase in the world population totaling 7.7 billion. In the medium variant, which appears to be the most plausible, by 2050 the world's population will have increased by 9.1 billion. The population of the industrial countries would stagnate at 1.2

billion, while the population in the developing countries would rise from 5.3 to 7.8 billion, which equals an increase of 50%.

In order to assess the socio-economic developments and the effect that population growth will have on the environment, the most important question needing to be asked is whether the population growth is more likely to be urban, or rural. Figure 2 shows that in the developing countries population growth will only be taking place in the cities, while population levels stagnate in the rural areas. In the industrial countries there will only be a slight rural decline, and respective urban increase in population figures.

Urban life is much more materialistic than rural life, because more buildings, traffic systems, and other infrastructures are built. This statement gains even more meaning when added to predictions of increased urban populations in developing countries.

A more detailed account of population growth is presented by Rainer Münz and Albert F. Reiterer in *Overcrowded World? Global Population and International Migration* in this series.

Increased raw material extractions and a continued rise in the emissions of pollutants

As producers and consumers, economized mankind leaves an imprint on the quality of the immediate environment as a result of the extraction of raw materials and the emission of pollutants. Up until now we have discussed the projections for expected economic growth and population developments, which already imply an increase in the extraction of raw materials and emissions of pollutants. However, technical progress can optimize the amount of materials and energy needed by manufacturers and should be taken into consideration, as well as an increase in private consumer's awareness of environmental aspects. How can these highly complex correlations be integrated into projections of environmental damage and linked to forecasts of economic development and population growth?

This can only be achieved by way of detailed models capable of illustrating the links between economic developments and environmental use, and developments at the level of a national economy's separate industries and commodity groups. For example, the extraction of the fossil fuels, oil, coal, and gas, and the pollutants emitted during incineration are all linked to the use of cars, the heating of homes, production of steel, and the generation and consumption of electricity. The extraction of gravel is important for the construction of buildings and roads, and the extraction of metals is particularly significant for other commodities and activities. Therefore, a detailed categorization of an economy's technologies, and the behavior of investors, producers, consumers, and even governmental conduct are required. Furthermore, this needs to be applied to all of the major countries, and foreign trade links between the international economies must be categorized into commodity groups. Studies conducted in the past, observing the behavior of consumers, producers, investors, and governments, can be converted into mathematical equations by way of expedient statistical methods.

GINFORS (Global Interindustry Forecasting System) is a model capable of achieving all of this. The model was developed by the Institute of Economic Structures Research (GWS) in Osnabrück, and has already been widely employed in research projects for the European Commission and German ministries. When presented with a population forecast, this model is capable of calculating economic developments for fifty countries, as well as the global extractions of raw materials and CO_2 emissions.

Together with Christian Lutz and Marc Ingo Wolter, I have used this model to compile a business-as-usual forecast: it assumes that aside from current policies, no further global environmental measures will be taken. To do this we used the previously mentioned middle variant of the United Nation's population forecast. We also set the model to give a close replica of the International Energy Agency's predictions of the GDP developments in various countries, which have also been mentioned at the beginning of this chapter. What state will the world be in if mankind isn't able to reverse the process of the utilization of natural assets?

Table 5 summarizes the results of global resource consumption, as compared with population growth. The world population will grow annually by 1.1%, the increase in the consumption of raw materials will be in places considerably higher, which will bring a steady rise to per capita consumption. The model calculates that the highest annual growth rate will be in the extraction of metals from the environment, at 3.5%. This can be explained by the increase in the capital stocks of machines, and other equipment, in the newly industrialized countries such as China and India. It must, however, be emphasized that these calculations include a reduction of material utilized per production unit resulting from technological progress. The same applies to the more efficient use of energy. Nevertheless there will still be a dramatic increase in the consumption of oil, gas, and coal. There will also be an annual increase by 2.4% in the use of non-metallic minerals, which mainly consist of building materials. The total outcome is an average annual increase in raw material consumption by 2.4%. This means that by the year 2050, we will be consuming 50% more raw materials than we are currently consuming today.

With a look to the emission of pollutants, and viewed against the backdrop of climate change, the carbon dioxide

	Growth rates in percent
Biomass	1.5
Coal	1.6
Petroleum	2.4
Natural gas	2.1
Ore	3.5
Other non-metallic minerals	2.4
Total extractions	2.2
Population	1.1

Table 5 Average annual population growth rate and global material
utilization from 2002 to 2020
GINFORS base prognosis
Source: Lutz, C., Meyer, B., Wolter, M.I. (2009)

(CO_2) emissions resulting from the combustion of the fossil fuels (coal, gas and petroleum) are of prime interest. Carbon dioxide, methane and four further atmospheric gases permit the sun's ultraviolet light to pass into our atmosphere and they also reflect the Earth's long-wave heat radiation. This is why we have an average global temperature of fifteen degrees, and not minus eighteen degrees, which would be the case if we did not have any carbon dioxide in our atmosphere. If concentrations of CO_2 and other greenhouse gases increase, there will also be a rise in global temperatures – which is referred to as the greenhouse effect.

Table 6 shows a summary of CO_2 emissions, in millions of tons, as emitted by countries or regions. In 2002 the United States were responsible for approximately a quarter of global CO_2 emissions. Even though the USA will emit an increase in emissions by 1.5 % by the year 2020, there will be a considerable decrease in the share of America's emissions equaling 21.3 % of global emissions. European emissions are expected to stagnate, which means a reduction from 16.1 % of 2002 to 11.1 % in 2020. The dynamics of developments in global CO_2 emissions will undoubtedly be

| | 2002 | | 2020 | | Average |
	million t	in %	million t	in %	annual growth rate 2002/2020
USA	5731	24.7	7439	21.3	1.5
EU-25	3739	16.1	3872	11.1	0.2
Japan	1144	4.9	1564	4.9	1.8
China	3381	14.5	5254	15.1	2.5
India	1054	4.5	1939	5.6	3.4
Other countries	8197	35.3	14818	42.0	3.3
World	23246	100.0	34886	100.0	2.2

Table 6 CO_2 emissions, in millions of tons, according to countries
GINFORS base prognosis
Source: Lutz, C., Meyer, B., Wolter, M.I. (2009)

governed by the newly industrialized countries China and India, where annual growth rates of 2.4% and 3.4% are expected. The remaining countries will also see similar average growth rates.

In terms of global CO_2 emissions this means an annual growth rate of 2.2%. The level of CO_2 emissions in 2020 will be 50% more than the levels in 2002. Consequently, the concentration of CO_2 in the atmosphere will continue to increase. Climatologists are expecting that already present concentrations of climate gases will raise the average temperature by approximately 2 degrees.

The business-as-usual forecast we have discussed is not at all exaggerated. On the contrary, one is more likely to assume that the emissions have been underestimated. Notice that in China, e.g., an average growth rate of 6.1% per year is expected between 2004 and 2020. Compared to this, an annual increase in CO_2 emissions of 2.5% is relatively small. Because China will be increasing its use of coal – the energy source with the highest

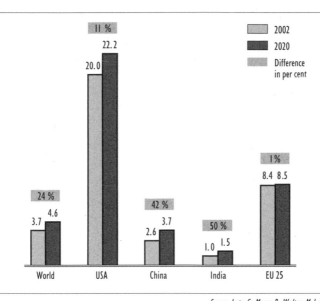

Source: Lutz, C., Meyer, B., Wolter, M. I. (2007)

Figure 3   The development of CO_2 emissions in tons per capita
GINFORS Base prognosis

amount of carbons – as its primary source of energy, its efficient use of energy will increase annually by 3.6%. Therefore the forecast implies that China will be taking considerable measures to improve energy efficiency. Added to this, the forecast expects a weakening in China's economic growth from 10% in the last 15 years to an expected 6.1%. It also becomes clear that the global average forecast is somewhat cautious: from 2002 to 2020 the gross world product will grow by 3.8%, but CO_2 emissions will "only" rise by 2.2%.

Figure 3 compares the development of CO_2 emissions per capita. To begin with, it is evident that an American emits twenty times as much CO_2 as an Indian, and approximately eight times as much as a Chinese person and 2.5 times as much as a European. By 2020, the global average per capita emissions will have increased by a quarter, in Europe they will stagnate, while in India and China there will be a strong increase, and in the USA a smaller rise in rates. The gap between America's per capita emissions and the global average will only have decreased from a factor of 5.4 to 4.8.

Tougher international competition for increasingly scarcer resources

By 2020 both China and India combined will be using as much energy as the United States, and all three of these countries put together will consume a total of more than 40% of global primary energies. A new competitive situation is evident on the international market for raw materials, especially with the advance of China and India, and also the Southeast Asian market economies. Recent years have already shown that a shortage in oil and metal supplies has brought a dramatic increase in the prices of these raw materials.

Competition is the essence of free market economic development processes, so in this respect there is nothing wrong with tougher competition. However, this only applies in an ideal context, where working markets are embedded in a stable legal system which, for example, prevents market dominating practices imposing themselves on supply and demand. Particularly the supply of metals and oil is often bound to regions that are politically unstable. One only has to think of the Middle East, Central Asia, Central Africa, and South America. There is a big temptation for the countries needing these raw materials to gain power over these markets by way of methods ranging from exerting political influence to engaging the military. There is a danger that countries become entangled in such conflicts, although it is also possible that large companies fuel up local conflicts in order to enforce their interests in these occasional legal black holes.

Will we experience outright resource wars in the future? This is exactly the criticism of America's involvement in Iraq. On the other hand, the employment of military power to secure supplies is widely accepted.

The repercussions of ever increasing environmental damage

At long last it would appear that the general public has come to acknowledge at least one central problem: climate change. This is due partly to the warmest winter in Europe on weather records, with more frequent and heavier storms. Around the same time, the media covered a report, by the English economist Nicholas Stern, on the correlation between economic development and climate change, as well as the fourth report from the Inter Governmental Panel on Climate Change (IPCC) which provoked heated debates. The IPCC consists of a group of 100 scientists that have been periodically presenting their reports to the United Nations for many years. Nicholas Stern is the head of a team of scientists commissioned to work for the British government. The Stern Review probably received so much attention due to Mr. Stern's tenure as the Senior Vice President of the World Bank, and in this respect – being a high-profile expert – his work can not be suspected as being 'fanciful.' Both teams of scientists have evaluated the vast amounts of literature on this subject, taking economic and natural scientific correlations into consideration. They reached the conclusion that a rise in the global average temperature by two degrees is already unavoidable, and that it is immensely important to prevent further increases. Yet this means that the concentration of greenhouse gases in the atmosphere cannot rise above 550 ppm. Consequently, long term CO_2 levels may not be higher than the amount plants are able to absorb, which is the equivalent of 20% of current emissions levels. If, then, a rise in temperature above two degrees is to be avoided, emissions will have to be reduced drastically. Because the climate system has such a delayed reaction, this does not have to happen overnight. But the longer until steps are taken, the more drastic the reduction measures will have to be.

Climatologists have discussed various alternatives. If the pinnacle of emissions is reached by 2015, then a yearly reduction of 1% would be enough to meet targets. If there is no change in direction before 2030, CO_2 emissions will have to be cut yearly by 4%. Occasionally such statements are interpreted to mean that we still have plenty of time to take action, and that we can wait for the major technological breakthroughs which will enable us to save energy considerably faster and more cheaply. But, why should these breakthroughs come along in 25 years, if in the meantime absolutely no efforts are made to change behavior as far as energy consumption and pollutant emissions are concerned? What kind of incentives will lead to the immense investments needed? After all, committing to new technologies has to be worthwhile. Let's take the introduction of hydrogen technology as an example. The fuel cell receives its energy from the connection of hydrogen and oxygen, and the waste material it emits is water. This system could replace present fossil fuel powered engines, and does not add any pollutants to the environment. Automobile manufacturers have already developed many prototypes which are very close to being ready for everyday use. The only obstacles to be overcome before the breakthrough of this innovative technology have little to do with the expected higher price of these vehicles, and more to do with the availability of a suitable network of filling stations, as well as a plentiful supply of hydrogen. Hydrogen, on the other hand, is obtained from water using the process of electrolysis, which requires large amounts of electricity. Obviously this electricity has to be produced by renewable energies, such as wind or solar power, because if the extra amounts of electricity needed were generated predominantly by fossil fuels, the use of the new technology would not result in lower CO_2 emissions. Car manufacturers will only start making considerable investments needed to make hydrogen vehicles when they are certain that a network of suitable filling stations will be in place, and that an ample amount of hydrogen will be available at a competitive price. The operators of filling stations will only make investments, reaching into billions, to set up and provide a network of filling stations if it becomes apparent that there will be sufficient demand for hydrogen. In turn, the electricity companies will only construct solar power plants and hydrogen electrolysis plants in the Sahara or Spain, for example, and build pipelines or special tankers to transport the hydrogen, once the demand for hydrogen is sufficiently high enough. How is all of this going to develop over the next ten or twenty years if energy policies do not set a course now to reduce CO_2 emissions? If we continue to wait, in twenty years time we will not be that much more technologically developed than we are today. Investors need clear signals.

Furthermore, with a global average economic growth of 4%, we would need from 2030 an 8% gain in annual productivity over a longer period of time in connection with the continued use of fossil fuels. This appears to be a hopeless venture. There is only one alternative: the turning point for global CO_2 emissions has to be reached by 2015. That leaves just seven years – not long considering that there will have to be worldwide consensus over targets and the necessary measures to ensure success. This means that a serious climate policy needs putting in place immediately. And even this is an ambitious course since, with an annual global economic growth of 4%, it requires increasing the efficient use of fossil fuels by 5%. Admittedly, consistent and stringent climate protection policies will cause a decline in growth; however, as we shall see later, these will represent relatively minimal losses.

This represents the greatest challenge to mankind in the 21st century. But even if this difficult undertaking is mastered, we will still be faced with substantial environmental damage as a result of irreversible climate changes, which will have considerable repercussions for human existence. Storms will become more frequent and more intense, glaciers will continue to melt, and the sea level will rise. Catastrophic floods will become more frequent and stronger. The oceans will become more acidic, which will have adverse consequences on fish stocks. A decline in Africa's crops is expected. The Amazon Rainforest will suffer serious damage caused by drought. The effects on the ecological systems will threaten the existence of 15–40% of the world's species. Human mortality will rise as a result of heat stress; malaria and dengue fever will spread.

The costs resulting from damages caused by extreme weather conditions, such as hurricanes, typhoons, floods, droughts, and heat waves have been estimated to rise, by the middle of the century, from 0.5 to 1.0% of the gross domestic product. As a reminder: the storms in Germany in January 2007 had, according to insurance companies, caused damages totaling approximately one billion euros. Claudia Kemfert from the German Institute of Economic Research (DIW) in Berlin estimates that damages from climate change in Germany alone will amount to 120 billion euros by the year 2025.

If we cannot manage to turn the tables on energy consumption and CO_2 emissions, and continue instead with the course forecast in Table 6, the results will be unforeseeable. Climate warming will probably have climbed from three degrees to five degrees by the end of the century, a development which will quite simply be catastrophic. To estimate the impact of such developments, it helps to realize that the difference between today's temperatures and temperatures during the ice age is a mere five degrees.

A detailed discussion on climate problems can be found Mojib Latif's book *Climate Change: The Point of No Return* in this series.

The continued rise in the population expected up to 2050 in developing countries will result in an increase of approximately 50% more inhabitants there by comparison with today's populations – who will need to be fed.

In view of climate concerns, it is unlikely that we shall see an increase in usable agricultural land, because this would involve uprooting the forests that we need to reduce the concentration of CO_2 in the atmosphere. It is also feared that, due to climate change, there will be an increase of desolate areas and steppes in developing countries, which will lead to a likely reduction of agriculturally productive land. Conflicts over whether agricultural land should be used for food crops, or to grow energy sources are also possible. There will be an increase in efforts to spread the use of biomass for energy production because biomass initially extracts CO_2 from the atmosphere before releasing it again when incinerated. In this respect it is neutral. The use of wood for heating, and ethanol as fuel does not increase the atmospheric concentration of CO_2. Boosting agricultural production with the use of more chemicals should be avoided due to the negative effect on the quality of soil and natural water sources. The biochemist Klaus Hahlbrock (*Feeding the Planet: Environmental Protection through Sustainable Agriculture*) fears that, without a more responsible application of genetic engineering, we will not be able to feed the world's population. Obviously this would also entail taking risks.

Our forecast on resource consumption estimated that by 2020 an average increase of the use of various materials would be 50%. There would be an exponential rise in yearly encroachments on the environment, causing considerable problems: first of all it is questionable whether future generations will have enough raw materials at their disposal. Furthermore, depleted mines could become a major burden, as can be seen today in Germany's Ruhrgebiet. Coal mining has caused a subsidence of the entire surface covering the Ruhrgebiet. Water has to be continuously pumped away in order to prevent the Ruhrgebiet from flooding. The federal state of North Rhein-Westphalia and the federal government have just reached agreements on who will bear the so-called perpetual costs. A further aspect involved in the mining of raw materials is the destruction of ecosystems, and consequently an accelerated extinction of species.

The central factor in the use of raw materials is that it plays a pivotal role in an economy's energy consumption. The processing of raw materials during the multiple stages of production requires the use of energy, as the example of iron demonstrates: iron is made by smelting ore, which is then rolled or poured involving further high amounts of energy. Finally, semi-finished products, such as machine parts, are made. In the next stage, e.g. in the automobile industry, the final product is turned out after an energy intensive mechanical assembly procedure. Obviously, materials are also transported between production stages, causing extra traffic and energy expenditure; the heavier the materials used, the more energy needed. This is the reason why

Friedrich Schmidt-Bleek (*The Earth: Natural Resources and Human Intervention*) demands a consequent dematerialization of production. Sustainable development is only achievable by halving the global consumption of resources. If we continue to let things drift along, an enormous gap will emerge between developments and targets. Bearing this in mind, Schmidt-Bleek's call for industrialized countries to increase the efficiency of their resource consumption by the factor ten is commendable.

Perspective future developments in human civilization are contradictory, and the results look gloomy. On the one hand, the ongoing positive economic developments in many newly industrialized countries will progressively spread to much poorer countries. This lends hope to the prospects of reducing poverty and hardship in the third world countries, which could eventually slow down population growth. On the other hand, improvements in economic conditions involve a heavy increase in the resources consumed by the developing countries and newly industrialized countries, as well as their emissions of pollutants. This will also be the case with important industrial countries such as the USA, and is the reason why economic growth is often denounced as being the source of all evil. Then again, this is the only way to improve the economic and social conditions of the third world. It is unrealistic to hope that this could be achieved by redistributing the wealth of industrial countries. It is easier to make sure every one gets their fair share if the cake is bigger. Moreover, it is thoroughly absurd to assume that it is possible to freeze the growth process. Consequently, we need to drastically separate economic growth from resource consumption. If we fail to solve this problem, then the climate changes that have already started will take on inestimable dimensions, which will have unimaginable consequences for mankind's existence on this planet.

3 What are the Causes, and What Kind of Solutions do We Have?

The difference between the personal and social costs of using the environment

Nobody can be excluded from using the environment, at least not by today's ethical and moral standards. This means that we can all consume the environment for free. Consequently, a rationally motivated individual is unlikely to be willing to pay the price for these assets by, e.g., abstaining from driving a car. This is what we call the "free rider" problem. Environmental assets are typically public goods without a market. As a result we treat nature as if it were available in unlimited amounts and unlimited quality. As economists would put it, we view nature to be a free good – like sand on a beach – even though this isn't at all the case. For ethical reasons we find changing our behavior difficult, because the damages we inflict on nature are generally quite negligible. They only develop as a result of so many people. What difference does it make then, if I'm the only one using environmentally friendly practices? Besides, these damages will only start affecting us some time in the future, and probably won't affect my life. This profile of the climate problem prevents many from behaving in a more environmentally aware manner, and necessitates political measures to be taken.

A commodity that doesn't cost anything is used more than other goods, which eventually leads to environmental damage.

The private cost of using the environment is zero, but the cost to society is enormous due to its scale and consequent deterioration of quality. Let's take the example of air: we use air as a disposal site for pollutants and greenhouse gases for free, and by doing so create the climate changes that lead to the known costs to society. Economists speak in this context of "externalities," because individuals don't take these costs into consideration when making budgetary decisions.

The imbalance between the cost to the individual and to society is the reason why nature is consumed so extravagantly. But then it is also clear that the problem can be solved by ensuring that individuals factor in the cost of using the nature when making decisions. This is also called the internalization of externality. Consequently, if there is an increase in the utilization of nature, and the quality of the environment suffers as a result, then the price for consuming the environment needs to rise, which would then reduce the extent to which nature is used. The end result is that the price for consuming the environment prompts more environmentally compatible behavior.

Economic instruments: tax, trading in licenses and subsidies

But how can we impose a price on the use of nature? Just the thought of it makes ecologist's hair stand on edge. Comments are made such as: "Economists know the price of everything, but the value of nothing!" The concerns of these critics is that the cause of our environmental problems is businesses striving to make profits, and unrestricted consumer consumption. And now they want to "economize" the environment – by putting a price tag on it?

But this is exactly the path we should be taking. Economists are convinced that the reason the environment is being damaged is due to a lack of market-based intervention, and not the opposite. If there was a price put on the use of the environment which reflected the scarcity of resources, it would affect the prices of all other assets, since resources are directly and indirectly included in each and every production stage in the value chain. So our present economic system is running on false prices, which is the reason why it is not able to treat nature more respectfully. The guiding principle behind setting the price of using the environment must adhere to the "polluter pays" principle, passing on the costs of using the environment to the individual, and in doing so create some form of conformity between the individual and social costs of using the environment. There are two possible ways of doing this. Either governments organize a competitive process, or they impose taxes on environmental use, which would bring about corrections in commodity prices.

The latter variation, the so-called "eco-tax" originates from the economist Pigou who, around eighty years ago, suggested the internalization of externality with the introduction of suitable taxes. Taxing the perpetrators leads them to reduce their activities, and the agents affected could receive compensation from the extra revenue. To put it simply, whoever extracts raw materials or adds pollutants to the environment will have to pay extra taxes. Having to pay taxes will immediately make the agent reduce the use of natural resources. Even more important is the indirect impact this will have: if, for example, a lignite coalpit has to pay a tax for extracting coal, then lignite becomes more expensive, along with the electricity generated from it. Depending on how much electricity other companies need, the price of electricity will raise production costs, leading to more expensive commodity prices, depending on the various production methods used. Eventually the more expensive prices of electricity and various other commodities reaches the private households. Private households and companies will react to increased prices by cutting their demand, which will finally have repercussions on the lignite industry, causing a decline in production.

Obviously the next question is how high the tax rate would have to be in order to reach specific ecological targets? Some people completely overestimate the problem. There are environmental economic models that give quite detailed predictions of the economic reactions to the tax, and also the effects these reactions would have on ecological targets. Besides, environmental policies generally tend to have long-term goals, advancing gradually towards the set targets. So in the process there is, naturally, time to make any necessary corrections to the tax rates. There is one further question concerning the use of the tax's revenue. It could be used to reduce national debt, or to reduce general taxes, or – as with the so-called eco-tax – reduce social security contributions.

Imposing taxes intervenes directly in the price system, and the demand for goods and the environment react accordingly. Fifty years ago the economist Coase favored the opposite solution, where governments set limits on the use of the environment, letting prices adjust accordingly. Governments create a market for environment utilization in order to create a shortage of environmental assets. To do this, the government first needs to set the amount of utilization of natural resources that it will allow. In the case of CO_2 emissions, this means that the government is willing to permit a certain amount of total emissions per year.

There are two variants for setting up a market process: in the trading solution, the government sells emissions allowances on a stock market. The competition between buyers would set the price per ton of emissions allowances, which would be payable to the government. The higher the demand, the higher the price will be. The companies that buy these emission allowances will have higher production costs, causing them to raise their selling price. Electricity suppliers will have to buy the necessary emission allowances on the stock market in order to cover the CO_2 emissions arising from their incineration of coal or gas; the extra costs would be added to the price of electricity. In turn, the more expensive electricity is, the higher the energy costs in the car manufacturing industry, or in engineering, and all other industries. Eventually consumers end up having to pay higher prices, which would match exactly the price of the CO_2 emissions caused by consumer demand. So if a government wants to cut CO_2 emissions, it only has to cut the supply of emission allowances available on the market. As a consequence, the price of so-called emissions certificates will rise, as will the price of all goods based on direct, or indirect influence on emissions. This will then cause a reaction in consumer demand. Households will reduce the consumption of the more expensive goods, the manufacturing companies will try to replace the more expensive primary products, or to cut costs by using different technologies. In any case, the price of emissions certificates will continue to rise until the reaction in demand of companies and consumers, and the demand for emissions allowances has been lowered enough to match the availability as stipulated by the government. In this way it would be possible to regulate natural resource utilization thanks to a government organized market. But how would the government's gained revenue be allocated? It could be used to repay debts, but it could also be used to relieve the burden of general taxes and essential expenses. In the latter case the money collected would be ploughed back into the economic cycle, stabilizing employment and wage developments in the process, which have naturally been affected by the inflation of costs.

However, some economists do not like it at all when

governments gain revenue, and they call for the organization of a different market for environmental utilization regulations – the so-called "Grandfathering": first of all the government awards all environmental users an initial set of user certificates for free, which would generally reflect past levels of environmental utilization. In the following years the government can then reduce overall utilization by canceling a certain percentage of the allocated rights. Let us turn again to the example of CO_2 emissions certificates. Those companies not able to reduce their emissions have to buy certificates from other companies, which means they have even higher costs. The companies that sell their certificates have adapted their production process, which brings higher abatement costs, whereby it pays for these companies to stretch out production as long as the additional abatement costs are lower than the price of the certificates. The marginal abatement costs (the altered abatement cost of the last production unit) balance out the price for the certificates.

How can we evaluate both of these variants? With grandfathering the first problem that needs solving is the allocation of the initial set of certificates. With the introduction in the EU of CO_2 emissions trading to the manufacturing industry, it became noticeable in recent years that there are considerable practical problems in creating a National Allocation Plan. After all, the main concern isn't only the allocation of emission allowances, but also the distribution of wealth offered by the chance to sell allowances. The amount of assets reflects the extent of damage to the environment made by the companies concerned. The advantage is that there is no redistribution of income between the government and industries, whereas with the auction variant there is a redistribution between both the industries involved and the government, and ultimately the government decides how it will use the revenue it receives from the trading of certificates.

But if we have trust in the government and its agents, this argument will be of little importance.

In both instruments the government regulates environmental use, the regulation reflects the supply of rights of use. These could be allowances for the emissions of pollutants, or also for the extraction of raw materials. The economic system adapts to these specifications by changing its prices and the consequent reaction of the demand curve. In the economic system, rights of utilization markets are like any other asset markets where speculative processes are practiced. So the price of certificates can be subject to considerable fluctuations, and does not necessarily follow a fixed course. This could pose added risks for companies that have to make long-term investment decisions with regard to increasing efficiency, and which could have a negative effect on targets being met. These complications are non-existent with taxes.

Taxes and rights of use provide incentives for companies and consumers to avoid polluting the environment. Of course, one can also create incentives to improve the quality of the environment by encouraging the development of technologies to replace the less efficient technologies. This is where subsidies are useful – but they are not really a particularly popular instrument with economists because they are often abused. It can make sense to support companies harnessing a more resource efficient technology, although maybe not yet cost effective, if it is likely one day to be competitively viable. Obviously the quality of an instrument depends heavily on the reliability of predictions. In Germany, renewable energy plants such as wind energy plants are subsidized by the government. The reliability of predictions here is relatively high, because electricity generated from fossil fuels will in the long term become more expensive, due to shrinking supplies of oil and gas, and also as a result of the indispensable climate protection measures.

Theoretically one can imagine an economy perfectly equipped with taxes and emission rights to help reach targets, where anyone extracting or importing raw materials to or from the environment, as well as anyone emitting pollutants into the environment, will have to obtain allowances from the government. Additionally one would have to consider how to impose similar allowances on the extraction of raw materials and pollution emissions of imported goods produced abroad. The latter is necessary, because otherwise producers of domestic products would consider their products to be subject to discrimination, especially when compared to imported products. Even if this problem could be solved with a global uniform system, or an equivalent taxation of goods at customs, or even by not considering foreign trade goods, there would still be further complications.

When illustrating the principles of regulation of allowances, we pointed out that indirect impacts play an important role in how a system operates. Let us examine the example of the market for CO_2 emission allowances. The price of CO_2 emissions rises, which makes energy consumption more expensive on every level of production, and for private households. In order for the system to function, it is crucial that price changes lead to changes in demand, and that CO_2-intensive goods are replaced by less CO_2-intensive goods. Once this happens, the burden of costs on households and companies will be relatively low, and there will be an actual and noticeable decline in CO_2 emissions. If the economic system is not this flexible, the economic advance performance structures and patterns of household consumption remain relatively fixed, which would require that CO_2 prices and the price of goods would have to respond vigorously in order to meet the same ecological target. The high costs to companies and the high loss in households' real earnings could challenge the acceptance of this measure. We have every cause to believe that this could be the case. Let us not fool ourselves – economic systems do not always run perfectly. There are market shortcomings arising from a lack of information available for companies and consumers, leaving them unaware of all of the options they have at their disposal when making economic decisions. On the other hand, there are competitive restrictions in many markets, which can hinder the necessary assimilations. Each and every pollutant and raw material will have to be looked at carefully in order to determine the candidates that will need integrating into a certificates scheme.

Let us imagine that present European emissions regulations for the manufacturing industry were extended to include transportation. In the eyes of the industry, fifty euros per certificate (or allowance) would be relatively high, whereby a private household might react relatively inflexibly. A car may emit two tons of CO_2 a year, which amounts to one hundred euros per household, or the price of two tank-fulls a year. Is this likely to make a household change its behavior? Probably not! So the industries are going to have to pay for stubborn consumer behavior with higher CO_2 prices. A further problem with taxes and environmental certificates can be found in the difficulties of distribution policies. A high-income household will be more able to pay for more expensive goods, whereas a low-income household would definitely be much harder hit. This especially applies to essential goods, such as heating. This is another reason why, when drawing up instrumental measures, a detailed examination needs to be made to identify the extent to which each category of goods would be affected, and to consider compensation for the socially deprived when necessary.

Taxes and environmental licenses are undeniably important tools for an effective environmental policy, but we cannot assume that they are the only measures necessary to meet targets. On the contrary, they need supplementary measures.

The necessary supplementary measures: communications and information policies, and cooperative solutions

As we were discussing economic instruments, we stressed that the economic process takes place on less than perfect markets. In fact, both companies as well as consumers are unaware of all of the options they have at their disposal, and they are also not completely aware of the ecological consequences arising from their activities. Furthermore, competitive markets are often restricted because certain companies are able to impose market-dominating practices. This is hugely detrimental to the effectiveness of economic instruments. Consequently, it can also be helpful for governments to improve the market participant's access to information, and also to take measures to improve communication between market participants.

It is consumers above all who often feel overwhelmed by the ecological and economic aspects when deciding which durable goods to buy, such as washing machines, refrigerators, heating systems, and cars. Generally, the decisive factors tend to be the price and the perceived quality of the product as reflected by a brand's particular image. In most cases, energy costs do not enter into the equation. Therefore, an economic instrument that would end up making these running costs more expensive would be ineffective. The government could achieve considerably better results by stipulating that product labels incorporate running costs. To some extent manufacturers are already doing this on a voluntary basis. Refrigerators and washing machines already have such labels, bringing home the message of the ongoing, running energy costs.

There is a widespread practice in companies to use controlling, an operational, managerial tool, to provide very detailed information on labor costs, but only minimal information – if at all – on material costs, despite the fact that the latter tends to be the most important cost component. This leads at times, particularly in smaller in companies, to almost grotesquely wrong decisions being made. Here again, the use of economic instruments will not make much difference. The way things stand, especially smaller firms would benefit immensely from the organization of consultancy agencies staffed with highly competent specialists. Such a facility already exists for example in Germany, the Effizienz-Agentur (efficiency agency) NRW, and has been working very efficiently for years. A similar institution has also recently been organized on the federal level, but has not yet been active on a large scale.

Encouraging companies to coordinate is especially important with regard to implementing technical improvements. The definitive solution to climate problems is technical development, in the form of new production methods that will save resources, and the development of new consumer goods that entail the use of fewer resources. It is significant that in the various sectors of the manufacturing and processing industries that are essential for Germany with regard to the development of new production methods (process innovation), the all-important factor for creating technical improvements and developments is cooperation between the respective branch of the machine building sector and the manufacturing companies. If cooperation is insufficient, or even non-existent, technical development within these companies will suffer as a consequence. Naturally, the state will not be able to advance technical developments on its own, but it has many channels at its disposal to foster cooperation activities. The state can play the role of a moderator, for example by organizing symposiums with efficiency agencies, as already mentioned, or the exchange processes between engineers.

Why does it always have to be the state that initiates each

measure? Obviously it makes sense for companies to push things forward in situations where the social consensus is all for minimizing the burden on the environment. Some companies are undoubtedly better informed than the government because, for example, they are better equipped to estimate the market potential of technological developments. In this respect it might be to the advantage of companies to anticipate governmental regulation and tax changes. Negotiations would then be made with the government in which companies from a specific industrial sector agree to reach set targets – e.g., reduce emissions – and the government agrees not to take any further measures in that sector before the year in which the targets have to be met.

The difficulty with such agreements for the government is that it cannot always tell if the offer made by an industrial sector reaches beyond the technical developments which, due to previous introductions of technical developments and long-term investments, would have been achieved in any case. If, on the other hand, the targets set are too demanding, it is possible that they might not be met at all. Because these agreements generally tend to be legally non-binding, there would not be any consequences initially. But the industry would naturally suffer considerable political damage, as is currently the case in the present public debate over the automobile industry's violation of voluntary agreements made concerning their fleet's average pollutant emissions.

In order to avoid this, during the initial negotiations, governments can threaten to implement measures, should promised targets not be met.

Alternative regulatory policy instruments

Regulatory policy instruments are derivatives of legal reasoning, comprising mainly of principles and prohibition. The core difference between regulatory policy instruments and economic instruments is that the individual is given little room to maneuver, but is told what is or is not allowed. On the other hand, with economic instruments the state merely gives incentives letting the individual make his or her own decisions, assuming that the individual will act selfishly and follow the direction desired by the state. Often regulatory policy measures are technical standards that companies and households need to adhere to. A salient example would be the introduction of Germany's extensive regulations for the Ordinance on Large Combustion Plants to reduce sulfur emissions. In this context the state has to set standards which companies are able to meet, although they should not set too low, otherwise the ecological target will not be met. In a dynamic setting this would mean the government has to estimate foreseeable technical developments. Some might doubt whether the government is capable of doing this. It certainly worked in the case of the above mentioned regulations. Within a few years the power plants in Germany had been desulfurized. But of course we have no way of knowing if an economic solution might not have been quicker and more efficient. In any case the central problem for regulatory policy is that, generally, tentative standards will be set to avoid the risks of them not being met.

One intelligent variation of regulatory policy is based on the technical standards set by the most efficient companies, and uses the so-called "best practice" technology. The remaining companies within the same industry are then set a deadline of a few years to meet this standard. Once this has been achieved, the whole procedure starts anew. Japan has had a great deal of success with this variation, which has been called the "Top Runner" program. With economic policy – such as a tax to reduce pollutant emissions – companies are given an incentive to find new technological solutions. This has a major advantage over every variation of regulatory policy. Competition can trigger the search for better technical developments that can exceed all expectations.

A further problem with regulatory policy is the enormous amount of bureaucracy required. On the one hand this entails considerable governmental efforts in setting up, for example, an authorization process. It also involves considerable costs for the companies that, for example, have to deal with long waiting periods for investment objects, and accept added risks in the course of their decision making.

In short, regulatory policies entail considerable extra costs, and there are doubts as to their efficiency. The main advantage is that they definitely bring about changes, which is the reason why they will always be favored over other instruments, especially for particularly critical situations, that might be life-threatening, or involve health hazards. Moreover, regulatory policy measures are easier to enforce because they are easily understandable for both lawyers and technicians, and the extra costs for those involved are not immediately identifiable.

Promoting intrinsic motivation

Regulatory policy instruments enforce changes in behavior, while economic instruments create incentives towards a more environmentally friendly behavior. But do we really need to be forced or seduced into protecting the environment? Of course we can also act of our own accord, as a result of ethical conviction, and without the need of external incentives. Psychologists call this intrinsic motivation. The issue here is our environmental awareness which is, more than anything, influenced by our education. This is the reason why environmental policy needs to recognize the importance of education. On all levels of general education, lessons either need supplementing with extra courses, or the material needs integrating in the existing curriculum.

Furthermore, the general public needs informing by the media of the consequences of environmental destruction, and also of opportunities we have to prevent further damage. This book is dedicated to this goal.

As a result consumers need to be informed of the effects their actions are having on the environment and, understanding the need for a more environmentally compatible behavior, make decisions, even though there might not be any direct economic advantages from doing so, or despite not having much choice in the matter. Ideally, conscientious and responsible business men and women will not only feel obliged to meet targets for short-term maximization of company resources, they will want to reach beyond and develop long-term company strategies involving environmental protection measures.

Ecological social market economy

There is a long tradition in Germany of discussions on the necessity for regulatory economic supplementation to safeguard social balance. Walter Eucken's representation of ordoliberalism is based on the very questionable assumption that full employment ensues provided that monetary stability is ensured by politics. At the same time Alfred Müller-Armack called for a deliberate economic growth policy and a trade cycle policy that guarantees full employment within the framework of a free enterprise system. The Social Market Economy was incorporated into the economic constitution in 1948 by Ludwig Erhard during the foundation of the German Federal Republic. The economy is driven by competition, which ultimately leads to the process of social selection. The winner performs best, the remaining competition is defeated. Because we all have different intellectual and physical qualities, obviously some market participants might always be the underdogs, no matter how hard they try. The market economy may be efficient, but there are no mechanisms to prevent such an unjust outcome. Consequently, additional regulations must be in place in order to improve the poor market results of the permanent underdogs.

The German Federal Republic's economic constitution has a marked wealth of state regulations to ensure social balance. They start in the fiscal sector where, for example, the state can use a progressive income tax to redistribute wealth from the high performers to the low performers, and also correct the distribution of market created incomes with numerous subsidies. Furthermore, there are also considerable restrictions in contractual liberty, which are designed to protect the socially disadvantaged in particular contracts. The various rent restriction regulations are a salient example here. Other restrictions have specific goals in mind, such as family security or the acquisition of new assets. Therefore, the use of economic and regulatory policy instruments to reach socio-political targets determines the character of our economic constitution. The promotion of intrinsic motivation is also nothing new, as can be seen in Ludwig Erhard's repeated appeals to industrial leaders for a sense of social responsibility.

It is only a matter of expanding the register of targets to include environmental policy goals, which obviously then need to be reflected officially in the constitution. From this perspective, the increased use of environmental policy measures should be viewed as further developing the social market economy into an ecologically social market economy.

This does not merely entail a verbal dimension, but is more about defining content. A competitively organized decentralized market form of economy is always at the core, in which companies strive to maximize profits and households maximize their utility. These systems create technical innovation and enable the accumulation of capital and thereby growth, as intended by economic agents. If we choose an ecological social market economy for our regulatory framework, we then lose the option of saving the world by lowering overall consumer levels. It then becomes necessary to separate economic growth from the use of resources. Basically this should be possible, because the price system, the technologies, and the quality of goods in an ecological social market economy would be completely different to those of today. We need to be wary of statements such as: "economic growth and the pursuit of profit are destroying the environment." Admittedly this statement applies to current events, but it would be wrong to consider it universally valid. An ecological social market economy has a completely different price system, and completely different consumer structures and technologies from those known today. It is only a question of whether we will be able to organize such a perfectly functioning market, or if we will have to arrange for a broader use of further instruments due to inevitable market imperfections.

The brief look we took in this chapter of various instruments showed that we cannot restrict ourselves to just one instrument. In particular we have to dismiss the notion that all we have to do is set up an extensive tax system and trade system of environmental rights of use. Measures such as coordination instruments, informative instruments, cooperation instruments, as well as regulatory policy standards, and the promotion of intrinsic motivation need to be added. The ensuing search for the right mixture will vary according to the type of resources extracted and pollutants emitted.

Policy makers on their own are overtaxed with creating a fitting context. There is a great need for extensive scientific support due to the highly complex interconnections. An excellent empirical data base would also represent a precondition, since speculation alone will not solve the problems and only the use of empirically valid hypotheses lead to workable solutions. The United Nations set up its System of Economic-Environmental Accounting (SEEA), a convention created to build economic environmental databases. The Environmental-Economic Accounting of the German Federal Statistical Office is based on it.

4 The Sustainability Paradigm

Now that we have determined our present position from two viewpoints – we have determined the Earth's present course, and where it will lead if no changes are made (Chapter 2), and we have looked at the possibilities of changing this course (Chapter 3) – all we have to clarify is which course we should follow.

The spirit of Rio

As early as 1983, during a United Nation's General Assembly, the former Norwegian Prime Minister Gro Harlem Brundtland was appointed the task of drafting a global program for change and demanding targets for the global community. In 1987 the World Commission on Environment and Development, chaired by Brundtland, presented the results in a report. This report defines a sustainable development as being a development which meets present needs, without endangering the needs of future generations. The term "sustainability" has been the central factor of environmental discussions for the past twenty years, and has been given various interpretations and comments. More than anything, efforts were made to define sustainability, in order to give politicians clear targets. In 1992 the report of the Brundtland Commission dominated discussions at the United Nations Conference on Environment and Development (UNCED) in

Rio de Janeiro. The conference was attended by governmental representatives from 175 member states of the UN, and many non-governmental organizations. The conference was a major success: for the first time a global agreement was met on development targets, and the measures that should be taken to meet these targets.

The so-called Rio Declaration is a kind of constitution for environmental and development policy, and its four pages contain a preamble and twenty-seven principles. It is useful to take a look at this declaration, which was signed by all of the states of the world, because much of what is discussed today in the context of national and international environmental policy has already long been preconceived. Let us restrict the principles we shall look at to those that are more relevant to our subject matter.

First of all, it is established that mankind is central to sustainable development (Principle 1). This is referred to as an anthropocentric concept, and it is therefore not exclusively about the conservation of nature. Furthermore, in accordance with the Charter of the United Nations and the principles of international law, the state has the sovereign right to exploit their own resources (Principle 2). Principle 3 reemphasizes balancing the interests of both present and future generations. Principle 4 calls for environmental policies which are also development policies. The elimination of poverty is declared to be an indispensable requirement for sustainable development (Principle 5). Special priority shall be given in developmental policy to the countries least developed and environmentally vulnerable (Principle 6). States shall cooperate in the spirit of partnership. The developed states recognize their responsibility in view of the technological and financial resources they have at their disposal (Principle 7). All states should reduce non-sustainable production and consumer patterns and promote sustainable demographic policies

(Principle 8). The exchange and distribution of innovative tech-nologies should be promoted (Principle 9). The states shall facil-itate and encourage public awareness of environmental issues (Principle 10). The states will pass environmental legislation and standards of stewardship and management. In developing countries these might be inappropriate and lead to social costs (Principle 11). The states will promote an open international economic system which will lead to economic growth and sus-tainable development in all countries (Principle 12). The states shall cooperate to prevent the relocation and transferal of activi-ties and substances that cause severe environmental degradation (Principle 14). A preventive approach is to be applied to protect the environment. Where there are threats of serious or irrevers-ible damage, lack of full scientific certainty shall not be used as a reason for postponing cost-effective measures to prevent environmental degradation (Principle 15). The national authori-ties should endeavor to promote the internalization of economic costs and the use of economic instruments so that the polluter bears the costs (Principle 16). Environmental impact assessments shall be undertaken for proposed activities which are likely to have an adverse impact on the environment (Principle 17).

Alongside the Rio Declaration, the legendary Agenda 21 – which presents a developmental and environmental policy action program for the 21st century – was also passed at the conference. The agenda comprises 359 pages, divided into forty chapters. The Agenda 21 initially addresses international organizations and the governments of each country. Under the motto "Think global, act local," it is debated that many global problems can also be solved through local activities, which is why the imple-mentation of Agenda 21 is also a task of communities.

The Rio Declaration has occasionally been criticized for not being specific enough to bring about immediate action. This is admittedly true, but it is also an exaggerated claim. The "spirit of Rio" had turned into a spirit of optimism, which had at least spread to all countries. Even if only letters of intent had been handed over, several milestones have been set for further developments in international environmental policy which cannot be simply swept aside.

The following proved, once again, to be the most important points: an anthropocentric approach will be taken, and therefore, the main concern is mankind's wellbeing. Environmental policy must always be viewed in relation to development policy, which means that the developmental needs of poorer countries have to be safeguarded. In the process an open international economic system should be promoted. Environmental policy should be oriented to a preventive approach, which also demands action to be taken even if the final certainty of serious environmental damage is not yet given. The various environmental policy instruments are cited and economic instruments are explicitly included.

The three dimensions of sustainability

The Brundtland Report portrays the decisive goal of sustainable development as its ability to permanently secure the needs of mankind. Right from the outset a global concept has been defined that extends to all future generations. In the process there should be an equal distribution of resources and goods spread across a generation and between the generations. This is an ethical postulate that needs no further explanation. If this postulate were continually ignored, it would not be possible to secure everyone's needs, which would lead to social conflicts. This highlights the social dimension of sustainability, which is not only present within individual generations (intragenerational), but also between generations (intergenerational). The academic debate over a clear definition of sustainability has given rise to sixty different variations, which we obviously will not be addressing in great detail here. We are more interested in the following main cornerstones of the debate.

The ecological dimension of sustainability is obvious, more than anything because of the intergenerational fairness in the satisfaction of needs: provisions have to be made so that all future generations have enough resources at their disposal in order to satisfy their needs. Not only does this include the quality and quantity of raw materials essential for producing consumer goods, but also the chance to continue using our environment in the same way that is possible today. This applies to biodiversity as much as it does to the functionality of the ecosystem, and the quality of water and air.

The economic dimension of sustainability arises not only from the needs for natural resources to satisfy demands, but also the needs for man-made consumer goods. We are talking about produced goods that should also be available for future generations to satisfy their needs. The central theme of economic sustainability revolves around the availability of the production factor capital, which consists of infrastructure – such as transportation networks and harbor facilities – and also fixed capital such as buildings and machinery. And of course, we should not forget so-called human capital – a term which has unjustly been labeled faux-pas word of the year by German philologists. Human capital is the term given to a society's body of knowledge embodied by its population. Sustainable economic management guarantees that this capital stock will be passed on to following generations.

We can give a somewhat more concrete and detailed definition of the term sustainability against this backdrop. Both the ecological as well as the economic dimensions of sustainability are obviously concerned with passing on a certain capital stock to the next generation, consisting of both natural stock as well as economic capital stock. Of course both the natural and economic capital should have a certain quality. It makes sense to presume that the quality and quantity of capital stocks passed on by each generation will at least represent the quality and quantity of the stocks they received from the previous generation.

It is slightly more difficult to apply the term "capital" to the social dimension of sustainability, since the term social sustainability only applies to a specific distribution of goods that are passed on. Yet the term "capital" has also found its way into this context. "Social capital" is the term given to the results achieved by specific institutions responsible for a community's social balance. This includes the regulatory legal framework of the social legislature, and the redistribution of wealth through the tax and social insurance systems, as well as negotiation practices between various representatives of different community groups. The presence of such institutions enables a certain level of social balance, and is subsequently called the social capital of a community.

One can imagine – admittedly somewhat abstractly – a sustainable development where one generation passes on its stock of social capital, natural capital, and economic capital to the next generation, in the same state it was in when they received it, i.e., of the same quality and quantity. The next question that arises is whether each of the three capitals are to be viewed separately, or whether the sum total of all three capitals is all that matters. This is where two concepts of sustainability come into play – "weak" sustainability and "strong" sustainability.

Weak sustainability is an interpretation by neo-classical economists, whose analyses are based on the ideal of "homo oeconomicus," who makes optimal decisions based on perfect information and steadily functioning markets. From this perspective one always prefers some leeway for economic decisions when posing environmental economic questions.

If the main concern is the sum of capital stocks, then as time passes one could replace one capital stock with another, for example natural capital could be replaced by economic capital. Obviously the advocates of this course are aware that there are limits to this kind of substitution. The difficulty is finding a uniform unit of measurement for capital stocks in order for them to be added up. Naturally economists think in monetary units. There are already calculations available on the economic capital for almost every country in the world. So 'merely' calculations for the remaining two capitals is all that is left to do.

Putting a price on nature is almost impossible. Maybe it might be possible on a local level to place an approximate value on the forest at the edge of town. One could ask town inhabitants how much they would be prepared to pay to use the forest. The responses could then be used to estimate the value of the local forest. This is a little harder on a national level. What value do rivers, forests, lakes, mountain ecosystems, and the seacoasts have to Germany's inhabitants? At the best one can estimate the value of a country's raw materials, which already have market prices.

Despite these difficulties, the World Bank makes regular estimates of the sustainable development of many countries based on the concept of weak sustainability. However, social capital is not included in the calculations, because it is even harder to calculate than natural capital. Economic capital is divided into human capital and the physical capital of machinery and buildings, which the World Bank deduces from a country's national savings. National savings consist of the remaining part of goods produced during the year, and which haven't been utilized, but added to the capital stock. Depreciation is then deducted from the various capital stocks. In the case of physical capital this would be the loss in value through wear and tear.

The deductions made to natural capital are based on the extraction of raw materials in a country and the environmental burden to nature. The extraction of coal, oil, and gas, as well as tin, gold, lead, zinc, iron, copper, nickel, silver, bauxite, and phosphates are all taken individually into consideration. Furthermore, the difference in forestry between tree replanting and depletion has also been included. Extractions have been valued at their respective market prices. The environmental damages taken into consideration are damages caused by CO_2 emissions, and also particulate emission damage. In 1995 the price per ton of CO_2 emissions was twenty dollars, the damages from particulate emissions equals the willingness, as ascertained in surveys, to pay for the prevention of such damages.

Since a national economy's national accounting does not consider the cost of education to be an investment in human capital, but rather as governmental as well as personal consumption, the costs of education are registered as direct accruals of human capital. The loss of human capital as a result of the loss in capital of those retiring during the year in question year has not been included in the study.

Table 7 illustrates the results of the World Bank's calculations, showing the examples of Angola, Brazil, Germany, India and the USA. The first column lists the total national account savings, the following columns show the depreciation in physical capital, in natural capital, education expenses, and the subsequent change in capital stocks in percent of each country's gross domestic product. According to these calculations, in 2004 Angola's development was unsustainable due to a dramatic decrease

	Savings	Depreciation of physical capital	Depreciation of natural capital	Human capital additions	Capital stock alterations
Angola	18.4	11.5	46.4	3.1	−36.4
Brazil	24.0	11.8	5.6	4.1	10.7
Germany	20.7	14.9	0.4	4.5	9.9
India	23.0	9.3	5.7	4.0	12.0
USA	13.4	12.2	2.0	4.8	4.0

Table 7   Capital stock alterations, according to the weak sustainability concept, in percent of the gross national income of selected countries during 2004

Source: World Bank, World Development Indicators 2006

in its national capital. All of the other countries appear, on the other hand, to paint a positive picture with positive changes to their capital stocks. Going by the results shown, we could call this a sustainable development, whereby the best figures are in India (12.0) and the poorest in the USA (4.0).

If we examine the results for all of the countries published by the World Bank, we see a picture similar to what Table 7 reflects. Generally, the industrial countries and the newly industrialized countries have such high growth rates in physical capital and in education expenses, that declines in the natural capital are overcompensated. Often the extraction of raw materials in these countries plays a small role, and is more prevalent in the developing countries, which leads to overall negative results in the developing countries. The overall global balance is, however, satisfactory and forms a stark contrast to our initial assessment of the present situation and expected developments.

First of all we need to realize that the extent to which environmental damages erode natural capital has been hopelessly underestimated in these calculations. Damages to natural water sources, tree population, and soil quality have not been included, neither have damages to the ecosystem, or to biodiversity resulting from the extinction of species. But also the price that has been set for CO_2 emissions, at twenty dollars a ton, is much too low. The damages arising from climate change are considerably higher.

Of course one can also adopt the more fundamental arguments made by ecological economists against the weak form of sustainability. We have already pointed out the difficulties of evaluating natural capital. For example, the prices fixed in the World Bank's calculations are simply arbitrary. The general objection to the concept of weak sustainability – as the examples show – is that there is a danger that economic capital will always have the upper hand, and that the demands for environmental protection become a farce.

The alternative to weak sustainability is strong sustainability, where substitution between industrial capital and natural capital is not permitted. In its most extreme form, not even substitutions within natural capital are allowed; instead it demands that each and every kind of natural capital is strictly preserved. This means that non-renewable resources – such as metals – should not be used at all. Renewable resources could only be used providing they can regenerate over a certain period of time. This would mean virtually excluding the extraction of the fossil fuels (coal, petroleum, and natural gas), since they are hardly likely to replenish themselves within a year. More liberal variations of strong sustainability allow substitution between the forms of natural capital. A slightly milder form of strong sustainability allows the substitution of non-renewable resources by renewable sources, whereby each designated use of the resources is viewed separately. This would allow areas to be covered by asphalt, providing other areas are then designated landscape conservation areas or nature reserves. At the same time it must be ensured that the environmental media's absorption rate for pollutants is not exceeded. This variation appears to be realizable and also ethical. The practical environmental policies of Germany and most European countries originate from this interpretation of sustainability.

Why some ecologists and economists do not like this concept

We have already seen that the successful introduction of the term sustainability by the Brundtland Commission still needs further clarification regarding the exchangeability between natural capital and industrial capital. As we have shown, the "mild" variation of strong sustainability (in which substitution between industrial capital and natural capital is not permitted, but is possible within natural capital) is the accepted compromised variation in practical environmental politics. As is always the case with compromises, a few are not at all pleased. A few economists are not happy with the rigid demands of ecological goals, where damages are measured in physical units – such as CO_2 emissions in tons – instead of in value units, which might then be deducted from other assets, as would be the case with the concept of weak sustainability.

On the other hand, some natural scientists are not happy with allowing substitution within natural capital, which in the long run has adverse effects on some environmental areas. In their opinion, this concession entered impermissibly into the discussion when economic goals were taken into consideration. They reject the triple bottom line concept of sustainability – ecology, economy, and societal aspects. They also occasionally criticize the anthropocentric approach, which places mankind's use of nature as being central.

The term sustainability has had quite a striking career. It

was originally used in communication among environmental researchers who, as we have seen, strived to find a clear definition for it. The extent to which sustainability has found its way into everyday use is a reflection of how the demands of environmental researchers are gaining recognition, and the measure of the public's awareness of environmental problem issues. It has become a fashionable word, used simply to express longevity. For example, somebody could make a sustainable impression on someone else. Some academics are not at all happy with this, which at times leads to an overreaction by completely avoiding the word all together.

From preventive environmentalism to sustainability strategies

Classical environmental protection regarded itself to be a repair shop, responsible for removing the dirt left behind in the atmosphere made by manufacturers and consumers. More than anything, it was concerned with installing filters to avoid emissions, for example, instead of changing our behavior. Of course, this kind of environmental policy was only marginally compatible with the sustainability concept. Arising from this concept of environmental politics, and at a time when most industrial countries were setting up their ministries for the environment during the eighties, it seemed to make sense to appoint this task to one authority. It was possible, to a great extent, to solve the problems independently of other governmental departments, due to environmental protection being considered secondary to the economy. Furthermore, mostly regulatory policy instruments were used and technical standards were set which, in view of their efficiency, could be represented autonomously by the environmental authorities.

In view of future tasks, the potential of classical environmental protection has been exhausted. Increasing resource productivity demands that technological targets are pursued, while
consumer behavior also plays a part. Environmental policies
must be integrated into economic processes. As we saw in the
previous chapter, it is also clear that aside from regulatory laws,
economic instruments and measures are required to increase
intrinsic motivation. Both the tasks that need executing, and
the instruments that should be used make it harder to separate
environmental policy from that of the other ministries. This
can be demonstrated with a few German examples: the task of
climate protection is assigned to the Federal Environment Ministry, whereas energy policy resides with the Federal Ministry of
Economy, because energy naturally plays an important role in
the economy. The Ministry of Economy is responsible for technology policies, which is a logical allocation, but – as we have
seen – it also has environmental and economic aspects too. There
are further tasks which overlap between the Ministry of Transport and the Ministry of Research.

Those were just examples of overlaps between various departments concerned with preserving the environment. However,
because sustainability contains ecological, as well as economical, and social dimensions, all political departments are involved
in creating policies. Against this backdrop it becomes clear that
creating a target-orientated and consistent sustainability policy
requires an enormous amount of coordination.

In Germany the task of solving this problem is the responsibility of the State Secretary Committee for Sustainable Development, chaired by the ministers of state in the Chancellor's
Office with the following departments as its members: foreign
office, finance, economics, consumer protection and agriculture,
employment, internal affairs, transport, environment, education

and research, health, economic cooperation and the family, senior citizens, women and youth. The work performed by the ministers of state in the Chancellor's Office is supported by the German Council for Sustainable Development. The members of the sustainability council, as it is sometimes called, are academics, and representatives of the churches, industry, unions, and environmental associations. The sustainability council sought out discussions with the general public using a wealth of events, including an online forum, and by the end of 2002 developed a 328 page long comprehensive strategy under the title "Perspectives for Germany." The council declares its belief in the triple bottom line model of sustainability, even though this concept has been controversial during consultations. Ecological, economic, and social sustainability targets were set generally for 2020, and 21 indicators were defined which enable us today to monitor developments and the annual progress made in meeting these goals. To this day, the Federal Statistical Office has already presented two reports on the respective balances of 2004 and 2006. Not only do the 21 indicators define each goal, they also describe so-called interim goals which, once met, will further advancements in reaching other goals. Consequently, improvements made in the energy productivity indicator simultaneously improve the values of emissions in the greenhouse gas indicator. The sustainability council also makes recommendations on the design of specific measures.

The complexities of coordination are multiplied by the fact that the member states of the European Union also run a joint sustainability policy. A collective EU environmental protection policy was initiated as early as 1970 in the fields of water pollution control, air pollution control, and waste management, for example – the typical domains of preventative environmental protection. In Gothenborg, June 2001, the Council of the

European Union committed itself to a sustainability strategy, which has been updated in many subsequent conferences. In March 2006 targets for climate protection were finalized under Germany's presidency. By the year 2020 the EU wants to increase the proportion of renewable energies to 20%, and at the same time also raise energy productivity by 20%. Up to the year 2020, CO_2 emissions are to be 20% lower than the levels in 1990.

The development prospects described at the outset are depressing: mankind is threatened by a climate disaster, which could lead not only to serious economic damages, but could also destroy the foundations of human life, if manufacturers and consumers do not alter their habits. In view of this situation we have no other choice. We need to revolutionize our technologies and lifestyles, which does not mean that we will have to forego prosperity. We must be open for, and want, change, and we will need an optimistic and pioneering spirit because, after all, society as a whole will not only have to tolerate the necessary changes, it will also need to actively participate. This requires the acceptance of a reorientation of technical developments and progress, which will place a greater emphasis on resource productivity. We will also have to be prepared to re-think our consumer habits and to give up some of the activities which we have grown fond of. These changes will not only happen as a result of the understanding of those who will be affected. Political measures will also be necessary in order to reach the targets that have been set. Basically, there is no alternative. In this respect, there is no room for pessimism.

Reconstructing the economy by increasing resource productivity

The result of the sustainability debate can be summarized as follows: the sustainable development of a society is given when its natural capital, economic capital, and its social capital remains intact, although the proportion of each capital may be different, however, there may not be any substitution between the three capital stocks.

It must be stressed again: sustainability is a normative concept that focuses on overall societal development. Although the standpoint of this book is concerned with the overall context, it singles out the question of how further environmental damages can be prevented by restructuring the economy. The decisive factor in reducing resource utilization is the national economy's consumption processes and its production of goods. We will start by considering economic development, because the dynamics of this subsystem are something that we must accept. Of course one could demand that economic growth should stop. It would make it easier to find the course leading to ecological sustainability. But economic growth is unavoidable in a dynamic competitive marketplace. Technical developments will be made, and the accrued profits will be used for new investments. But how then should national economies be structured? The alternative, centrally planned economies, has failed from both an economic as well as an ecological view.

The western industrial societies, whose lifestyles and economic organization is becoming increasingly adopted by the developing countries and newly industrialized countries, aim for an economic development in which adequate economic growth and full employment are achieved, while prices generally remain stable. As it is not easy to achieve all three goals at the same time, this is often called the "magic triangle" of economic policy.

If we take the mild variation of sustainability portrayed above, substitution is tolerated between the natural forms of capital, which then permits the use of raw materials. The focus on the depletion of resources then shifts on to how the environment is affected by pollution emissions as a result of using raw materials. The amount of emissions is limited by the amount the environment can absorb. The term "sink" is often used in this context. Let's take a look at the example of CO_2 emissions. The environment annually absorbs 20% of the CO_2 emission levels of 1990, mainly through plant photosynthesis. Sustainability requires, then, that the global level of annual CO_2 emissions be reduced to this amount. As long as emissions are above this level, the concentration of CO_2 in the atmosphere will continue to increase. The longer it takes to reach the absorption rate, the higher the rise in the Earth's temperature. Once we consult climatologists in order to set specific limits for global warming, the climatologists will be able to tell us which course we need to take, and which levels we need to abide by in order to reach the target set for a specific date.

Now the economists are called on again. How can we combine economic growth with the necessary reductions of CO_2 emissions? The following equation helps us find the link. The E to the left of the equal sign stands for emissions. On the right there is also an E, but

$$E = (E/R) \times (R/Y) \times (Y/B) \times B$$

it is fragmented in four separate expressions which are all multiplied together. After the multiplication has been done, the R in the first expression is cancelled down by the R in the second expression, the Y in the second is cancelled by the Y in the third, and the B in the third is cancelled by the B on the far right, which obviously leaves only an E to the right of the equal sign. The equation is then an identity. The R in the equation stands for resource utilization, the Y stands for gross domestic product, and the B represents the size of the population. Consequently

E/R equals emissions per unit of resources used, which in our example would be CO_2 emissions per energy unit. R/Y equals resource utilization per unit of gross domestic product, and Y/B equals gross domestic product per capita.

If E should fall, even though the population (B) and per capita income (Y/B) are growing globally, then the emissions per resource unit (E/R) and resource utilization per unit of gross domestic product (R/Y) will have to drastically decrease. So it is logically conceivable to have economic growth and an increase in population growth, as well as simultaneous reductions in emissions.

In order to define sustainability targets in a concrete manner, we will have to consider all pollutants. A public dialog, based on information on the consequences of pollutant emissions provided by natural scientists, will set a schedule for emission goals. An example of this are the present EU decisions to single-handedly reduce CO_2 emissions by the year 2020, by 20% of 1990's emissions and, should other states participate, to extend this further to 30%. The repercussions for emissions per resource utilization (E/R) and the resource utilization per unit of GDP (R/Y) should be deduced from the predictions for economic growth and population development. The so-called remedial environmental policy concentrates on the target dimension E/R and tries to minimize the amount of emissions per resource utilization. The old style of environmental policy assumed that mankind would behave as it always has, and that eventually sewage treatment plants, filters, and catalysts would collect pollutants. We do not want to completely belittle the significance of this policy. It has made essential achievements in keeping the air free of sulfur and other substances, as well as improving the quality of natural water sources in Germany – but it is unable to cope with solving future challenges. The potential of so-called end-of-pipe-technologies to reduce pollutant emissions has been substantially exhausted. The only interesting option is, however, offered by CO_2 sequestration during the incineration of fossil fuels (CCS: Carbon Capture and Storage). After sequestration, the gas should be stored deep below the Earth's surface, obviously making sure that it does not escape and rise back up to the surface. The first pilot plants are presently being put into operation. There are still plenty of technological difficulties, and with the present energy prices the technology is also not yet profitable. Experts are not expecting full implementation before 2020. If this technology were to be made available, it would solve the problems of electricity power generation, especially for countries with large reserves of coal – such as China.

Twenty years ago Friedrich Schmidt-Bleek, the former head of the Materials Flow Department (*Stoffströme*) of The Wuppertal Institute, declared that the definitive target dimension should be resource utilization per unit of the GDP. How this ratio develops all depends on which goods a country produces and uses. Let us stay with our example of CO_2 and energy: what share do transportation services hold in terms of consumption? Are cars the main form of transport, or are public transport systems used more often? Consequently, it is the structure of consumption which plays an important part. Furthermore, the amount of energy used per unit of GDP is also dependent on production technologies. How much energy do the vehicles and machines used need? How well insulated are the buildings? How much energy is required to manufacture the sheets of metal used to make vehicles, and which technologies are employed to make steel?

This example shows how various factors influence the ratio of resource utilization per unit of GDP. As a result the scale of this ratio needs slimming down. To do this, the reciprocal – GDP per resource unit – needs to increase. Let's stick with our case, which makes things more clear. It is a question of how many units of GDP can be created per unit of resources. This can also be called resource productivity. Increasing resource productivity should be the decisive aim of environmental policy measures, because it would automatically bring a reduction in emissions. This does not mean everything depends solely on the technological aspects. Our daily decisions concerning the kinds of goods we consume, and the decisions made by companies over which technologies to harness in manufacturing all influence developments in resource productivity.

What effect will this concept have on the economy? Obviously the government's objective is to speed up technical development, which will also bring both new consumer goods, as well as encourage new production processes that require less raw materials. Such developments always entail investments in new machinery, buildings and other assets. This affects the industrial sectors which play a particularly important role in the German national economy – as we have already seen in Chapter 2. If increasing resource productivity is the only plausible reaction to the challenges of climate change, then Germany is on the one hand an important agent able to develop and supply the necessary technologies, and on the other hand, it is a country that could economically profit from such developments.

5 What Options Are There For Increasing Resource Productivity?

In the previous section, we learned that in order to reach our sustainability targets, we require an integrated policy approach, which tries to influence the entire national economic structures with a view to increasing resource productivity. The following two chapters will attend to how our findings can be applied to present opportunities in view of the global challenges outlined in Chapter 2. The strategies have two different lines of attack: one is concerned with changing manufacturers' behavior, so that they use more materials efficient technologies, and the other is directed at consumers to change their lifestyles by using fewer resources. We will not be examining the necessary measures here just yet. First of all, we want to gauge the means we have at our disposal to increase resource productivity with current technologies. We would also like to try to estimate which technological developments would allow us to increase the use of these potentials in the future.

The Sufficiency Strategy: the role of the consumer

In the previous chapter we advocated the theory that it should also be possible to have sustainable development in connection with continuous economic growth. If, under this premise, we ask what are the potentials of a strategy that has a starting point of consumer behavior, it is clear that the sufficiency strategy does not expect us all to tighten our belts and use less. The focus is on consumer patterns, and not total consumption. Which are the commodities that we demand? Sufficiency means preservation by abstaining. This does not have to mean renouncing consumption; it merely applies to resource utilization.

Optimized resource utilization, as a result of changing consumer patterns, has been successfully achieved when there is a concentration of resource utilization for just a small group of goods. If we can then use less goods from this group, and more from other groups, we will be able to maintain the same level of overall consumption, and still lower the amounts of raw materials being used. The dynamic aspects are even more important: with higher concentrations of resource utilization for less categories of goods, the effects of product innovations which require lower amounts of raw materials will be much more effective in reaching the target than would be the case with equal distribution.

Martin Distelkamp, Marc Ingo Wolter, and myself conducted a study, for the Aachen Foundation Kathy Beys, investigating consumer spending, in the year 2000, of private households in Germany. We divided consumer demands into 43 categories of intended use. We calculated how much raw material utilization in Germany would be affected if expenditure on utilities were to fall by one billion euros. All direct and indirect connections, including the raw material extractions for goods imported from abroad, were taken into consideration. For example: reduced expenditure by private households on package holidays initially leads to a reduction in demand for the services of travel agency services and transportation, such as airlines and trains etc., and also the hotel and catering industry. The transport industry needs less fuel and electricity, which leads to a reduction in demands for the processing of mineral oil and generation of electricity, in turn leading to less demand for oil, gas and coal, some of which is extracted in Germany (coal) and some of which is extracted abroad (oil and gas). The catering industry prepares fewer meals, and as a result orders fewer primary products from the food industry, thereby lowering demand for goods from the agricultural industry. Eventually the agricultural industry uses less biomass. Plus, a considerable portion of package holidays consists of services abroad, which means the same calculations need to be made for the extraction of biomass and fossil energy fuels abroad.

If German private households reduce their expenditure on package holidays by one billion euros, there will be a reduction by 679,500 tons in raw materials extractions, of which almost a half, 328,600, can be accounted for abroad. Calculations were made separately for fossil fuels (oil, gas, coal), metals, industrial minerals, construction minerals (gravel, stones, sand), biomass, excavation and dredging, and erosion. Table 8 illustrates the results, listed according to the importance each intended use has on the total resource utilization.

The base expenditure of one billion euros helps compare the results. However, the proportion of each intended use in the total budget has not been included in this illustration. This applies in particular to the intended use "solid fuels including district heating" at the top of the list, with a reduction of 59,589,000 tons. In the year 2000, its total volume had been barely one billion euros. In this respect we are viewing a one-time-only total reduction of district heat and solid fuels. On the one hand this is due to private households' direct demand for coal, and on the other hand, it is due to the input of fossil fuels used to manufacture district heat and also the use of 2,826,000 tons of biomass. Because only minimal import levels have been accounted for, it

Position	Reduction in private consumption, according to intended utilization, by a billion €	Δ TMR in 1000 tons	Of which	
			Domestic material utilization	Imported materials
1	Solid fuels (incl. district heating)	−62964.4	−58099.2	−4865.1
2	Electricity	−28109.6	−25133.3	−2976.3
3	Garden produce etc.	−4383.2	−3540.6	−842.6
4	Glassware (among others)	−3241.1	−2361.7	−879.4
5	Installation / Repair of housing	−3215.7	−2512.8	−702.9
6	Foodstuffs	−3016.8	−2051.4	−965.4
7	Alcoholic beverages	−2896.5	−1929.2	−967.3
8	Non-alcoholic beverages	−2689.3	−1716.0	−973.4
9	Other consumer durables	−2403.7	−1125.2	−1278.5
10	Transportation services	−2046.3	−1207.4	−838.9
11	Household goods	−1998.0	−969.6	−1028.4
12	Catering services	−1927.2	−1145.9	−781.4
13	Accommodation services	−1912.3	−1134.3	−778.0
14	Gas (including liquid gas)	−1809.8	−872.8	−937.0
15	Footwear	−1804.2	−804.8	−999.5
16	Photography and computer equipment	−1799.4	−735.7	−1063.7
17	Installation / repair of private motor vehicles	−1656.9	−991.8	−665.1
18	Personal commodities	−1652.5	−907.5	−745.0
19	Purchase of vehicles	−1549.2	−329.2	−1219.9
20	Liquid fuels	−1482.7	−676.6	−806.0
21	Fuel	−1482.4	−676.4	−805.9
22	Tools and equipment	−1361.5	−413.1	−948.4
23	Furniture and the like	−1225.3	−521.9	−703.4
24	Personal hygiene	−1061.8	−493.4	−568.4
25	Newspapers, books, etc.	−1056.2	−418.4	−637.8
26	Water supplies, etc.	−1021.3	−801.2	−220.1
27	Medical products	−889.8	−322.4	−567.4
28	Housekeeping products and services	−852.3	−300.3	−552.0
29	Financial services	−840.7	−604.8	−235.9
30	Clothing	−826.4	−235.7	−590.7

Table 8 The results of cutting expenditure on resource utilization (TMR: Total Material Requirement) by a billion, in 1000 tons, for the most important categories of intended use in Germany, during 2000

Source: Distelkamp, M., Meyer, B., Wolter, M.I. (2005)

would seem that district heat is produced mainly using lignite coal.

Second on the list is electricity, which is the most predominant factor leading to the extraction of fossil fuels in Germany. A reduction of one billion euros equals a reduction of approximately 5% of the electricity used by private households. This is easily achievable with the use of more efficient housekeeping, such as by rigorously turning off electric appliances (i.e., turning appliances off completely, and not using the stand-by function), using energy-saving lightbulbs, and only using washing machines and dishwashers for full loads, all of which would not impinge on our standard of living. Yet the savings made in resource utilization amounts to 28,109,600 tons, which equals just 20% of the savings that could be achieved, if all 43 consumption categories would be reduced by one billion euros. If we were to leave solid fuels out of our calculations, due to their low percentage in the total budget of private households, then the one billion euros saved in electricity equals 30% of the total savings we have just calculated, which would require reducing total consumption by forty-two billion euros. This is also impressive in reference to the levels of consumption and material requirement in 2000: the observed drop in consumption of one billion euros is less than 0.9 per mille of consumption of German private households and companies. So we have now identified private households' demand for electricity as being the biggest influencing variable in Germany's resource utilization.

Trailing behind, in third place, are garden products with 4,383,000 tons, which comprise 1,732,700 tons of fossil fuels, 1,560,000 tons of biomass, 608,300 tons of erosion, and 249,600 tons of construction minerals. Glassware is in fourth place with 3,241,100 tons, of which 1,433,400 tons counts as construction minerals, 373,700 tons as metals, and 248,100 tons as industrial minerals. This recount is only to demonstrate that Table 8 has a background of a very comprehensive data of records.

If we add together the resource utilization of "transport services," "maintenance and repair of automobiles," "automobiles bought," and "fuel," we arrive at the total sum of private households' expenditure on mobility. Here, again, is a calculation that demonstrates clearly the comparative quantities. Just as we did for electricity, we calculate a reduction in demand by around 5%. So, yet again, hardly an order of magnitude that requires fundamental changes in behavior, but simply a matter of reducing the occasional superfluous excursion or two. Reducing the demand for mobility by 5% would, however, still equal eight billion euros. If we distribute this sum to match the configurations of 2000 in each intended use, the result would be a reduction of raw material utilization by 12,971,000 tons. These calculations show a fall in private consumption of 0.7%, and a reduction of raw material utilization in private households and industries by 0.2%. The noticeably weaker results, compared to electricity, obviously has to be given in the significance of lignite coal in the production of electricity in Germany. An important difference between both results is also due to the fact that changes in mobility affects approximately two-thirds of raw material extraction abroad, whereas with electricity this only the case with 10%.

Table 8 alone confirms hopes that the use of raw material really is concentrated to a select group of utilities and respective commodity groups. If we can reduce just the first ten intended uses by one billion euros each, this would give us 76.6% of the total reductions in resource utilization that could be achieved if we were to reduce all groups by one billion euros each. Changing consumer patterns would then have a considerable effect on overall resource utilization.

Table 8 also offers further important information. In some

groups a reduction of consumption in Germany would have more effect on resource extraction abroad than it would domestically. This is shown, e.g., with automobiles, which constituted 5.2% of total private consumption with 65.5 billion euros in 2005. Reducing private consumption by one billion euros would reduce raw material utilization by 1,549,200 tons, which would affect 1,219,900 tons from abroad. This is due, on the one hand, to the share of imported automobiles. On the other hand, metals account for 650,000 tons of the total reduction, which are obviously also completely imported for cars manufactured domestically.

These calculations are only to illustrate that significant results can be achieved on the utilization of resources by changing consumer patterns. This means that by changing consumer habits, even without technological changes to products, relevant results can still be achieved. Once this has been achieved, companies would then concentrate their innovations on those areas that would reap the most benefits from consumer demand.

The Efficiency Strategy: Factor 10 innovations and investments to increase resource productivity

In *The Earth: Natural Resources and Human Intervention*, in this series of books – and in a wealth of further publications – Friedrich Schmidt-Bleek has very clearly described a multitude of examples that clarify how resource utilization can be reduced by a factor of ten, as the name of the "International Factor 10 Club" suggests, of which he is president. The central focus is on technologies used in various areas of national economic processes. An additional study for the Aachen Foundation Kathy Beys, with Martin Distelkamp and Marc Ingo Wolter, and myself, attempted to identify the important branches and technologies for reducing resource consumption on a macro-economic level.

A national economy's technologies can be illustrated by comparing, on the one side, the use of the inputs labor, capital and intermediate products (materials and services produced by other firms), with, on the other side, output results of the firm in question. Output, is defined, from the point of view of a supplier, as being finished products, as well as intermediate products (materials and services from other companies and used in the production process). Finished products are goods demanded by private consumers, or capital goods which are added to capital stock, and also exports.

Intermediate products are interesting from the perspective of resource utilization, since it entails the actual physical utilization of resources. The use of intermediate products links the production of all industries and branches. In the disaggregation of fifty-nine industries, as defined by the German Federal Statistics Office's records of production patterns in their so-called input/output tables, each industry supplies the other industries with its intermediate products. This could also be called an intermediate product cycle.

Of the fifty-nine industries, only eight (agriculture, forestry, fisheries, coal mining and the peat industry, oil drilling and gas mining, uranium mining, ore mining, stone and earth) extract raw materials directly from the environment and feed them into the intermediate product cycle. Ore mines, for example, extract iron ore to supply the steel industry for it to make steel, which is then processed into metal goods that eventually become components in various capital and consumer goods. On each production stage the respective intermediate products receive additional labor and capital inputs. Not only are the raw materials introduced into the cycle by the eight chosen domestic extraction industries, they are also imported, which is especially important for Germany. When determining a country's use of raw materials, it additionally makes sense to include the indirect raw materials contained in imported finished products. In this context, Friedrich Schmidt-Bleek descriptively speaks of "rucksacks" of raw materials, brought into the country by way of the imported finished products. Therefore, we can calculate the sum of domestic and imported raw materials that are used when one production unit of an industry is manufactured.

We carried out two experiments in model calculations, taking into consideration the connections just described. In the first experiment we consecutively reduced the input by 1% in each of the fifty-nine industries, and calculated the effects this would have on macro-economic resource utilization. A 1% reduction of intermediate input might possibly be due to technical developments, but we will not be going into this here. Our focus is on the macro-economic effects of raw material utilization. The decisive factor in our analysis was that we could determine the direct and indirect macro-economic effects of increased efficiency within each sector by applying the input/output approach. For example, the utilization of all of the inputs in the automobile industry is reduced by 1%. Among other things, this would lead to a reduction in supplies from the steel industry to the automobile manufacturers, who would in turn use less coal and ore. This results in a reduction in demand for national coal, and demand for ore from abroad. A further possible effect assumes that the automobile manufacturers will use less synthetic materials, meaning a reduction in the amount of polymers processed, and ultimately fewer deliveries from the chemical industry to the plastics industry, not forgetting the reductions in the amount of energy and oil used in the chemical industry. Again this leads to smaller amounts of fossil fuels extractions both domestically

Position	Reduction of input coefficient production sectors by 1%	Δ TMR in 1000 tons	Of which	
			Domestic	Imported
1	Energy, etc.	−15,165.9	−13,568.6	−1597.3
2	Construction	−8625.8	−6949.8	−1676.0
3	Metals and semi-finished products	−7465.2	−1549.1	−5916.1
4	Foodstuffs and beverages	−6283.7	−4316.2	−1967.5
5	Glass, ceramic, etc.	−5743.7	−5149.9	−593.8
6	Automobiles and parts	−4375.7	−1465.5	−2910.3
7	Metal products	−3472.2	−866.0	−2606.2
8	Coal and peat	−3184.7	−1688.9	−1495.9
9	Chemical products	−3070.8	−1749.0	−1321.8
10	Machinery	−2688.0	−803.9	−1884.1
11	Coke products, mineral oil products	−2287.5	1193.3	−1094.2
12	Public administration services, etc.	−1445.5	−1109.6	−335.9
13	Paper, cardboard and goods	−1424.1	−704.8	−719.3
14	Agricultural and hunting products	−1416.4	−1117.5	−298.9
15	Property and housing services	−1379.4	−1144.9	−234.4
16	Electrical appliances	−1286.3	−405.5	−880.8
17	Accommodation and catering services	−1251.7	−775.6	−476.1
18	Retail services, etc.	−1122.6	−790.0	−332.6
19	Health and veterinarian services, etc.	−1121.6	−775.2	−346.5
20	Credit institution services	−947.0	−722.4	−224.5
21	Land transport and transportation services	−828.9	−605.4	−223.5
22	Trade middlemen / wholesale services	−810.3	−553.0	−257.3
23	Rubber and plastic products	−780.6	−477.9	−302.7
24	Wood; wood products, cork and wicker products	−762.4	−545.1	−217.2
25	Business relations services	−725.3	−488.8	−236.4
26	Publishing and printing services	−633.8	−279.2	−354.6
27	Automobile trade services, repairs, etc.	−620.9	−367.5	−253.4
28	News, radio, television and the like	−588.2	−204.1	−384.1
29	Education services	−522.6	−389.0	−133.7
30	Furniture, jewelry, toys and the like	−518.1	−214.6	−303.5

Table 9 The results of reducing all input coefficients within a production
sector by 1% of macro-economic raw material utilization
(TMR: Total Material Requirements) in 1000 tons, for the top
30 most important industries in Germany, 2000

Source: Distelkamp, M., Meyer, B., Wolter, M.I. (2005)

and abroad. These examples demonstrate that improving efficiency in one sector has an effect on the remaining sectors, and in the process influences the utilization of raw materials across the board.

Table 9 shows the results for the chosen industries, ranked according to the total amounts of raw materials used. Here again, calculations were made individually for fossil fuels, construction minerals, industrial minerals, biomass, excavation and dredging, and erosion, but omits showing the details. Naturally, after the results of private household consumption, we expect electricity to be at the top of the list. Here the utilization of fossil fuels would be directly reduced by increased efficiency. In second place we find the construction industry, where increased efficiency could directly reduce the use of minerals. With metals and semi-finished products we find the steel industry and non-ferrous metal industry (copper, aluminum) on third position, which would directly save metals and also fossil fuels through increased efficiency. Furthermore, electricity is used to process aluminum and copper, and reductions in electricity would also reduce the amount of fossil fuels used in the energy industry.

On the remaining places, we only find a few sectors of the extractive industry with noticeable direct effects, such as glass/ceramic, coal, and peat. The other important industries in the remaining top ten are capital goods manufacturers, such as automobile manufacturers, machine manufacturers and metal commodities, the chemical industry, and the nutrition and luxury foods industry.

Improving the efficiency of the top ten most important sectors by 1% already yields 59,566,000 tons, or a 70% of the total 1% reductions in raw material utilization from all of the fifty-nine industries combined. Approximately 37% of the reductions in raw material utilization would be made abroad. Efforts to attain technical developments to reduce raw material utilization would need then to concentrate on the primary industries and capital goods industries.

One might ask if the observations conducted, which are focused on the separate industries, are sufficiently detailed enough. Certainly not all of the input from one particular industry carries the same importance for the utilization of raw materials. Conversely, industries that have been identified as being of little importance for raw material utilization may contain separate significant input.

In order to identify this better in the second experiment, we reduced, one after the other (successively), each of the 59 × 59 = 3481 inputs (fifty-nine distributing and receiving industries) by a further 1% and separately calculated the effects on the macro-economic utilization of raw materials for fossil fuels, construction minerals, industrial minerals, biomass, excavation and dredging, and erosion. The results show that only a small portion of supply relationships play an important part in the use of resources.

Figure 4 depicts the forty most important technological interrelations in the context of total raw material utilization in Germany. The sectors are marked in boxes; the arrows running between them show the supplier relationships. Distributions are also made within industries, since an industry comprises many companies that can be interlinked via chains of distribution. This can often – as in the metal industry – be an indicator of the conglomeration of consecutive production stages. The metal industry comprises the manufacture of steel, non-ferrous metal, and copper, as well as processing sheet-metal and other semi-finished products. The thickness of the arrows represents the degree of direct and indirect importance in relation to the total economy's

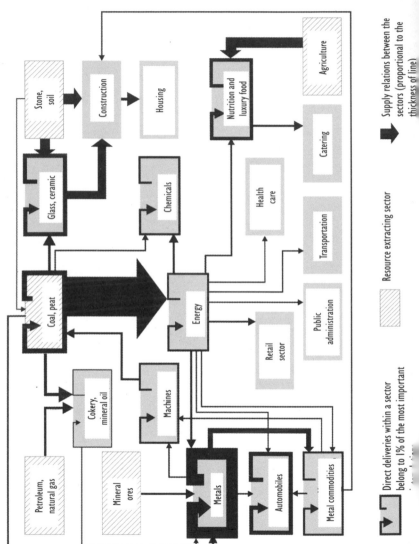

Supply relations between the sectors (proportional to the thickness of line)

Resource extracting sector

Direct deliveries within a sector belong to 1% of the most important utilization of raw materials. If the supply relationships portrayed (which only represent around 1% of total supply relationships) were to be reduced by 1%, it would equal two-thirds of the total 1% reductions of all 3,481 supply relationships.

Bear in mind that this illustration only portrays input linkage within an industry. This is the reason why, households, as recipients of electricity, have not been included.

Figure 4 identifies four importantly interlinked areas, sometimes called clusters. The most important cluster is the use of coal to generate electricity and to produce coke, which is then used to process metals. In this instance we are talking about the production of aluminum and copper, as well as in the electric arc steelmaking process. The metals are then used to produce cars, machines, and metal commodities. The third cluster is the construction industry group. This is where the intermediate forms of stone and soil, such as construction minerals, for example, are used in the final processed form of glass and ceramics. Steel is also an important component in the construction industry, which then links the construction cluster to the metal cluster. The cluster for nutrition and luxury foods is the center of biomass extraction. Agriculture extracts biomass and delivers it to the nutrition and luxury foods industry. Energy is a further important component here. The output is then transported to the catering industry.

Even though the service sectors – such as trade, transport, health, and public administration – are not quite on the same scale as the industrial clusters just identified, they still have significant energy needs.

We can summarize that the production of energy, and its distribution, obviously dominates material flow patterns. The roles

Figure 4 Technological interrelations and resource utilization

of the remaining clusters differ greatly in the economic production process. On the one hand we acknowledge both the metal manufacturers in the automotive industry, mechanical engineering, and metal commodities as being manufacturers of capital goods, and the chemical industries as partaking in the international competitive market, and predominantly exporting their products (Chapter 2). The two remaining clusters – construction/ stone, soil/housing, nutrition and luxury foods/gastronomy – on the other hand are noticeably pegged to the domestic market. These differences, which we shall cover in the next chapter, will play an important role when designing an efficiency policy and the choice of instruments to be applied.

The importance of key technologies

Under the direction of the physicist and philosopher Armin Grunwald, the Hermann von Helmholtz Association of German Research Centers published a comprehensive study, which competently sheds light on all aspects of Germany's sustainability problems. Among other things it was established that four key technologies are particularly promising for the more sustainable development of technologies. They are nanotechnology, biotechnology, renewable energy technology, and information and communication technology, and their further development will be of great importance in the design of fundamental production conditions for all economic sectors.

New technologies go through multiple stages before finally being introduced to businesses and households. The process of innovation starts first by basic research being made by universities and established research institutes, and also in corporate research departments. The next step involves transforming the findings into concrete product ideas and new production processes. The profitability of these developments plays an important role in this process. The final stage involves introducing the new products and production processes to the market. Consequently, the implementation of innovations always entails a combination of research of both natural science technologies and the economy. The key technologies mentioned are presently at various stages within the innovation process. This is why statements of possible applications are more precise in some cases than others.

Nanotechnology is not strictly a specific technology in the production sense of the word. It is a more general advance of technology into the realm of atoms and molecules, and its potential was recognized as early as 1959 by the physicist Richard Feynman. In 1974 the Japanese Norio Taniguchi first defined it as being a technology to divide, join, and deform materials using atoms or molecules. If it is possible to work with materials on the level of atoms and molecules, then it is possible to give materials completely new properties. It also allows products and production methods to be scaled down to dimensions previously considered to be unimaginable, and which subsequently creates totally new application potentials. This is the reason why exponents of nanotechnology fondly speak of the third industrial revolution.

The development of new and advanced materials is exceptionally important for all industrial production processes. As we have seen, steel, aluminum, copper, and ceramics are all important industrial primary products, which make up a large percentage of material flow. If, for example, it is possible to reduce a material's weight, yet at the same time retain its solidity, it would enable previously inestimable reductions in energy expenditure if these materials were then used in the production of automobile and machines. It is expected that with the development of more efficient insulation materials and windowpanes with heating and cooling properties, unforeseeable energy savings in homes might also become a possibility. It is also possible that products and production methods could be constructed which had not been possible with previous materials. Developing machines on the nano level could offer completely new diagnosis methods and control technologies for many sectors, such as in medicine. Completely new methods for data storage, for example, would be available for information and communication technology, which is a powerful driving force behind technological progress. Nanotechnology opens up new possible applications for information and communication technology. Nanotechnology has already generated a wealth of new products, such as surface coatings that eliminate the need for material intensive cleaning methods, for example. The sunscreen lotions which protect against ultraviolet irradiation contain nano particles, and modern processors already contain nano-sized structures.

The term bioengineering was coined by the botanist and microbiologist Raoul H. Francé (*Plants as Inventors*, 1920). Its fundamental idea is centered on using nature as a model for technical solutions. Today the very dynamic scientific discipline bionics (sometimes called biomimetics or biomimicry) follows the same concept. In the German language the term bionics comes from biology and technics. A differentiation has to be made between bionics and biotechnology, which is mainly concerned with biochemical processes and their industrial conversion technologies. Yet another branch is genetic engineering, which covers all proceedings in the area of molecular biology and genetics used to alter an organism's genetic make-up.

The application areas of bionics are exceptionally diverse. The advanced materials for lightweight constructions are especially interesting in view of increasing resource productivity.

Nature has been optimizing the principle of lightweight construction with the evolutionary developments of plants for millions of years. These principles of construction can be adapted and applied to industrial manufacturing.

Another example of the application of bionics involves the wet ability of surfaces when droplets of water fall and hit the surface, which is of great importance for the technologies and efficient processes in many industries. In this process, hydromechanics plays a very decisive role in the content of the drop of water, and its ultimate effect. Nature offers interesting methods of influencing wet ability. One example is how the shape of the lotus leaf repels liquids.

Genetic engineering, and in this case, especially green genetic engineering, which deals with altering the genetic constitution of plants, offers opportunities for more sustainable development. Research is focused on breeding plants resistant to pathogenic germs, which consequently leads to lower doses of pesticides being needed. Even more importantly, it is concerned with species able to offer bigger harvests, which could play an important role in fighting widespread global hunger. The use of genetic engineering is very controversial. Obviously, for each case any potential risks need to be scrutinized impartially and scientifically. Generalizations are counterproductive, and could lead to opportunities being missed.

Renewable energies have been an important pillar in energy supplies for centuries. Corn was ground in the Middle Ages by windmills and watermills, and in Holland it would not have been possible to gain land by draining the dykes without the power of windmills. In the early days of industrialization, metal could only be processed where watermills were powered by valley rivers – as in the Siegerland and Sauerland regions of Germany, for example. In more recent times windmills and watermills have become more idyllic relics of "the good old days," and had been replaced everywhere by the more powerful fossil fuel powered plants. It was not until the first oil crisis in 1973, and the debates on the limits to growth, which were started around the same time by the Club of Rome and others, that this technology came back into remembrance.

Renewable energies either directly use solar power, wind or wave power, or thermal energy from deep below the Earth's surface. The incineration of biomass is also considered to be a form of renewable energy. All of these processes characteristically do not produce any CO_2 emissions. The same applies to biomass combustion, because the plants burned only contain as much carbon dioxide as they absorbed from the atmosphere when they were growing.

There is a multitude of technologies generating renewable energies already in operation: dams store hydro power, and the water outflow powers the generators, thereby producing electricity. Tidal power plants use ocean currents on the coast – triggered by the moon – to drive the turbines. Wind turbines generate electricity from the movement of their blades. Semi-conducting materials in solar cells create electricity through the sun's radiation. In solar thermal power plants, sunlight is concentrated by numerous lightweight mirrors and used to create steam, which then drives the turbines. There are plans, for particularly sunny parts of the Earth, to use this clean form of electricity to create hydrogen out of water using the procedure of electrolysis. This could then be transported in ships, or through pipelines, for further use as fuel in automobiles. Its combustion in car engines would be "clean," and the only emissions would simply be a few drops of water. Prototypes of hydrogen-powered cars are already being tested.

The physicist Joachim Nitsch expects that Germany's

electricity consumption can be completely covered by renewable energies, whereby his calculations include electricity imported from Mediterranean solar thermal plants. The technologies that have been developed to date are not yet able to meet the competition of those using fossil fuels. Technical progress may usher in considerable changes in this respect. Above all, the steadily rising fuel prices caused by an increasing shortage of fossil fuels will benefit renewable energies.

Over the last twenty years information and communication technologies (ICT) have dramatically changed both global business related work practices, as well as consumer habits. Without computers, the Internet, and mobile phones, the economic developments of globalization would not have been possible. At the same time, the use of these technologies has contributed to a new consciousness, and created new lifestyles.

The effects of developments in raw material efficiency are inconclusive. Earlier predictions of the paperless office as a result of the advance of ICT have certainly not come to fruition. On the contrary, the limitless access to information, and the opportunity to print it out immediately, has given a new lease of life to those with collector tendencies. Electronic scrap and waste, which collects at a rapid pace, and on a considerable scale as a result of technological progress, has also created new problems. The same applies to the increase in electricity needed to operate ICT.

At the same time, the flows of materials have also been influenced positively: e-commerce, which still holds considerable untapped potentials, saves resources by cutting down on delivery journeys, for example. The ability to send e-mails with attachments has revolutionized communication. Many trips for meetings have become unnecessary, and economic processes have as a whole seen a dramatic increase in efficiency. The electronic control of production processes has given birth to resource saving technologies. Telematic systems can be increasingly deployed in future to direct traffic in such a manner that resources will be saved. Environmental policies will step up its use of systems to collect fees, as is already the case with the road toll for trucks. The utilization of information and control systems can be harnessed for more efficient production processes for both agriculture and industry.

6 What Precisely Needs to Change to Enable Increased Resource Productivity in Europe?

Now that we have decided, in Chapter 4, on the course economic changes should follow in order to improve resource efficiency, and placed this course within the overall context of a sustainability policy, the next question that needs answering is: which measures should be taken? To answer this question, we will refer back to Chapter 5, where we concluded that selecting a suitable strategy covering the choices of technologies used by both consumers and companies offers considerable potential for sustainable development. Reducing the ten most raw material intensive, intended consumer uses by one billion euros (total consumption reduction ten billion euros), still employing current production technologies, would alone deliver three-quarters of the total reductions obtainable if reductions were made of a billion euros in each of all the categories of intended use, equaling a total forty-three billion euros. In companies, the concentration of technological clusters we have identified is even larger: raw material utilization is concentrated in energy generation, metal production and processing, the construction industry, and also food production. If we view the interdependence between all economic sectors, a reduction of 1% of the most raw material-intensive supplier relationships (using current technologies) would alone make up two-thirds of the total results achievable if all supplier relationships were reduced by a particular percentage. These results raise hopes that restructuring the economy is in the realms of the possible, since raw material utilization is concentrated within certain commodities and technologies. This notion gains momentum when looking at the potential dramatic reductions in raw material utilizations offered by the key technologies of nanotechnology, bioengineering, renewable energies, and information and communication technology. So the next questions that need answering are: how can we persuade the companies in the clusters using raw material intensive technologies to exploit these potentials in the future? How can we prompt consumers to consider the direct, and above all indirect raw materials contained in the products they buy daily? The question we shall be tackling in the next section concerns the instruments that would allow us to tap into more sustainable development processes. In this context, we will take a look at present regulations, and ask which kind of changes are necessary.

Sustainability is indivisible

Influencing the behavior of consumers, manufacturers, and investors through environmental policy measures will also have an effect on the economy and society. Taxing the use of electricity, for example, reduces the amount of electricity used and pollutants emitted, but in the process the state would also be taking income from consumers, and using it in some way or other. Such taxes are always more painful for low earners than they are for those in higher income brackets. These observations are trivial, but in the context of a normative concept that postulates three dimensions of sustainability – the ecological, the economic, and the social – they play a decisive role during the process of creating a policy. Obviously, a policy which halves consumption would dramatically improve ecological sustainability, but at the same time, it would cause an economic crash leading to catastrophic unemployment. Therefore, the economic and social dimensions of sustainability would deteriorate.

Competitively organized economies are characteristically growing economies, since dynamic competition generates technical development and profits, leading to new investments. Western industrial societies are now dependant on dynamic economic development, because they are facing tough competition from the newly industrialized countries that have long adopted the basic elements of western economic structures. For western industrial countries, including Europe and Germany, economic sustainability means the capability to stand up to competition. In Europe, the social dimension of sustainability requires, among other things, that there is full employment, and that the distribution of income does not adversely affect the lower classes. If we mean business with sustainability, then when creating our policy, we would be wise not to restrict our deliberations to increasing resource productivity, we should also address the effects measures would have on competitiveness, and how social balance would be influenced. In this respect it is not only environmental policy that is under close scrutiny, but also the entire structure of economic and social policy. We will not be able to go into great detail here, but we will have to discuss the most important topics. Restructuring the economy to increase resource productivity will only be possible in a dynamic and innovative economic environment, which on the other hand is only achievable when competitiveness and social consensus prevails. In this respect certain topics will have to be addressed which reach beyond the tight frame of resource management. Demographic change is of central significance, and the changes it will entail for the social system, education policy, and the labor market.

The role of economic instruments

Economists believe that the root cause of the environmental problems is that we are able to use nature for free, and because of this
we "misuse" it to such an extent that natural stocks are endan-
gered (see the detailed account in Chapter 3). We falsely treat
nature as if it were an unlimited resource, which is why its price is
set at zero. In reality, however, it is a limited good. Because natural
resources are contained, in one way or another, in the products
we produce and consume, the prices of the goods are erroneous.
Naturally, the economic decisions we make as consumers and
producers, based on these prices, are also wrong, and lead us ever
closer to the hardships of ecological catastrophe. From this perspective, the recommended therapy is quite simple: all we have to
do is assess the price of resource utilization based on its scarcity.
In order for this to happen, the state has to establish markets for
environmental commodities, and in the process limit the supply
of natural goods by the amount necessary for conservation goals
to be met. An example of this is the trading of CO_2 emissions
rights within the sector of European primary industry. An alternative would be that the state does not install such markets, but
instead corrects existing price systems by taxing resource utilization, or subsidizing resource saving technologies.

This would be a perfect solution to our problem, providing one
essential condition is always met, namely the functionality of our
goods markets. In Chapter 3 we took a detailed look at this aspect
and, to recapitulate briefly, discovered that unfortunately markets
are not always completely functional, which means that manufacturers and consumers do not always make decisions based on
price signals, as market participants are not always the perfectly
informed calculators they are presumed to be in economist's
theories. Furthermore, some markets are lacking in competition,

because there is too little supply or demand. A closer look needs to be taken of specific cases. Therefore there is a primacy of economic instruments, but they will not function without regulatory policies and appealing to the intrinsic motivation of the public.

Further developing emissions trading allowances

Against the backdrop of the commitments written into the Kyoto Protocol to reduce greenhouse emissions, on October 13 2003, the European Union decided to start up a trade system for greenhouse gas emissions (Emissions Trading Directive). The member states set up a so-called National Allocation Plan (NAP) for a limited period of time, which stipulates the total number of certificates they intend to distribute over that period of time, and how the certificates will be allocated to the companies. Emission trading is restricted to the extractive primary sector, such as the generation of electricity, iron and steel production, the paper and cardboard industry, stone and soil, glass and ceramic, as well as mineral oil processing, and coking plants. The allocation is based on each plant's latest emissions. In the process, the total number of allowances distributed by a country should to be based on the requirements stipulated in the Kyoto Protocol for 2008 to 2012, in order for these agreements to be met. The allocation ruling allows each country considerable discretionary powers, as they only have to commit themselves to a specific emissions target, without having to specify which industries will be subjected to these targets. The core of the EU regulation is intended to be a free allocation of allowances, although it also permits states to auction up to 10% of the total volume. The EU directive views greenhouse gas emissions as being trade objects, although only CO_2 emissions are involved during the first phases.

Trading began in 2005. The allowances are valid from 2005 to 2007, and in a further period from 2008 to 2012, also underpinned by the Kyoto Protocol. Further phases are planned for the future, consisting of five years each. After trading started, certificate prices reached peak values of thirty euros per ton of CO_2 – probably due to uncertainties over this new institution – only to fall to below one euro in 2007. In other words, the market collapsed, implying a too plentiful supply of allocations. The philosophy behind the market is that the industries needing more certificates than they have been allocated then take on the role of the demander, while those that have used less emissions supply the amount of certificates saved. The availability of certificates sets the price, which in turn gives an incentive to cut down on CO_2 emissions.

It is quite conceivable that many aspects of these instruments will be modified. First of all, it might be asked if further economic sectors, or even consumers, should become subject to emissions trading. Furthermore, it should also be examined whether distribution practices should stay as they are, and whether a larger amount of the overall supply, or even the aggregate supply, should be auctioned. Let us turn to the latter question first.

With the free distribution of emission allowances – also known as "grandfathering" – actual costs are only incurred when a company needs to purchase certificates. If the company's emissions are within a range that does not require the acquisition of extra certificates, then there are no further real costs. Yet the certificates still play a decisive role in a company's bidding decisions. Manufacturers use the certificates available, and forgo the option of selling their surplus. So the company passes up the opportunity of profit, which could also be interpreted as being a cost. Economists speak of opportunity costs. The extent to which these opportunity costs are considered in a company's price calculations depends on how tough competition is, and the price elasticity of demand – the reaction of demand to modified prices. In any case, the German electricity producers – of which there are four – have justified the dramatic increase in the price of electricity in 2005 and 2006 by claiming, among other things, that this is due to the opportunity costs of the certificate trade.

These kinds of price increases generate considerable profits for those companies able to mention opportunity costs in their price calculations, whereby the ultimate cause for increased profit is in the allocation of free certificates. Not enough experience has been had yet to tell the full extent these prices will effect further production stages, because the certificates market is currently (2008) not completely efficient. However, we can have a look at estimates of the two most extreme cases.

Let us take a look first at the case where all companies participating in the certificates trade completely mark up their opportunity costs. In the case of EU emissions trading, this means that we would see considerable increases here in Europe in the cost of electricity, steel, and ceramic, etc. For those companies that use these raw materials as primary products, there would be a sizeable increase in costs which, depending on the international competitive market situation, might not be completely transferable, leading to corresponding profit setbacks. Should international competition permit shifting these costs, it is ultimately the consumer who carries the burden. From an economic standpoint, there would be problems caused by international competition, from an ecological standpoint the result would appear to be initially desirable, as the goods which directly or indirectly contain energy input and raw materials such as steel, ceramic, etc., would become much more expensive in Europe, leading to a decline in demand. However, we should not forget that more of these products would then be produced abroad and imported by Europe, so the global ecological balance would do anything but improve. All that would remain would simply be a competitive disadvantage for the European economy.

In the second case, we assume that when calculating prices the companies participating in emissions trading only take into consideration the actual costs arising from the eventual additional purchase of emission allowances. The effects on prices would be considerably weaker, as only a small percentage of the total certificates stock would actually be traded. The subsequent cost pressure on companies further down the value chain is minimal, and there are unlikely to be any difficulties with competitive capacity on the international market. However, the desired ecological impact, namely diminishing demand for products heavy on raw materials, would be missing.

When the state auctions the certificates, companies have to pay full price for the certificates. In this respect we can expect the same steep increases in the price of basic commodities as those resulting from the shift of opportunity costs in the case of entirely freely allocated certificates. All other effects concerning successive production phases and also affecting consumers are also identical – including the desired ecological impact. The only difference is that now the state receives the proceeds from emissions trading. Of course, these proceeds should not be retained, but should be ploughed back into the economy. The respective literature discusses income tax reductions or cutting the contributions for social insurance. This would help the general economy, but the extractive industries would only indirectly profit from this. This would still result in redistribution from the primary industry to all other areas of production and private households, which in the long term leads to the relocation of these industries. There would be no ecological gains from this, because relocating production to the newly industrialized countries (where raw material efficiency is even lower) would in fact increase global CO_2 emissions. On the other hand in Germany, for example, added value would be lost. Furthermore, German industries, with their focus on manufacturing capital goods, definitely rely on basic commodities such as steel and ceramic. It is very important for further technological advancement that the entire value chain is as local as possible.

This is the reason – in my opinion – why it is necessary that the revenue from the certificates auctioned is given back directly to the industrial sectors. For example: companies in the ceramics industry use ovens to fire their products. They have to attain enough certificates to cover their CO_2 emissions. The total revenue from the certificates received by the state from the ceramics industry should be, in my opinion, accredited back to the companies in the ceramics industry, prorated either according to sales, or production units as a key. As a result, the sector as a whole remains unencumbered, but the individual companies with high CO_2 emissions have to pay, and the more efficient companies profit from provisions. The desired impact on CO_2 emissions still remains, because it still pays for the companies to improve their technologies. The industry as a whole would remain unburdened. So there would be an incentive at each level of production to substitute energy sources and for technical innovation, without incurring any expected adverse effects on international competitiveness. When reimbursing auction revenue to electricity manufacturers, obviously renewable energy providers, which have already benefited appropriately, should also be reimbursed. The new plants in this sector could then forgo assistance from the controversial Renewable Energy Sources Act, which we shall come back to later.

A future development in emission trading should then include converting to certificate auctions as soon as possible. However, as a precautionary trust-building measure, there ought to be regulations guaranteeing that the auction revenue will be fed back into the economy.

Another further development could consist of including additional industries in the certificates market. Although the question then arises of whether it is sensible to overlap various economic instruments. In Germany the so-called "eco-tax" on gasoline, diesel, heating oil, electricity and gas, came into effect in 1999, and with some exceptions, affects all companies and private households. The exceptions are generally the industries already targeted by emissions trading. These are the industries that use the most energy, and have to pay considerably lower actual tax rates. Up until now, kerosene has been affected neither by the eco-tax or emissions trading. So in this respect emission trading could be extended to include airlines, as is presently being discussed.

The previously mentioned argument against double burdens speaks against an inclusion of private households in emissions trading. However, mention should be made of the interesting suggestion by David Flemming that an extra emissions scheme should be created to cover private energy consumption. We will discover that this suggestion offers a genuine alternative to the taxation of private household's energy utilization.

Under the heading "Personal Carbon Trading," the Tyndall Centre for Climate Change Research made the following proposal: each consumer receives free CO_2 emission rights. The allocated permit is valid for a set period of time, and is stored on a card from which deductions are made when purchasing fuel, public transport tickets, gas, heating oil, and electricity. Consumers using less than their allocation are free to sell their surplus allowances. The technical requirements for such a system are more or less already available for fuels, since filling stations already have electrical booking systems for sales made using credit cards. The amendments necessary would be minimal. Those who want to avoid the use of cards can buy the allowances directly at the filling station.

Advocates of this system stress that this path leads to the

practice of a more conscious use of energy, which would strengthen rational consumer behavior. Furthermore, consumers directly learn the consequences of their behavior concerning CO_2 emissions. But above all, it is possible to control an important component of energy demand through the total amount of emission rights, without incurring any negative economic consequences. In addition, private carbon trading would be more preferable than an energy tax for households, as it would ensure a fairer allocation per capita. Accordingly, everyone could use energy without any additional costs, providing they stay within their per capita allowance. It is even possible to improve one's status by using less and selling allowances. Only those persons using excess amounts of energy will have to pay extra.

The ecological effects of emissions trading for private households are superior to the effects of energy taxation. Households generally react poorly to changes in the price they have to pay for the energy they use, which means that the demand for energy and also the household CO_2 emissions would only be minimally reduced. In fact, it is possible to control household emissions of CO_2 through the amount of total allocations. If it is not possible to implement personal carbon trading, then the private tax rates for electricity, gas, heating oil, and fuels will have to be raised.

The ecological tax reform

The heading "ecological tax reform" is used for an alteration to the entire tax system that alleviates the labor production factor, and burdens the "natural resources" production factor. With this reform, the Swiss economist Binswanger expects "double dividends" to result in the form of relief for the environment and increased employment. The energy taxation in the United

Kingdom, Sweden and Germany has been declared as being
an introductory ecological tax reform. In Germany it taxes the
energy consumption of fuel, heating oil, electricity, and gas, and
the revenue yielded is paid into the pension fund. The repayments
made from the ecological tax revenues alleviate the contributions
employers and employees pay into the pension's fund. Therefore,
this can genuinely be called an introductory ecological tax reform.
However, there are still a few blemishes, because the tax rate is not
orientated on the CO_2 emissions from energy sources, and also
because it is not a resource taxation only for industries. It is in
essence a burden on households, because the industries that have
particularly energy-intensive production methods have received
generous exemptions. If we were to replace energy taxation with
private carbon trading – as suggested in the previous section –
there is not much substance left of the so-called eco-tax.

Consideration could also be given to enhancing the ecological
tax reform by taxing the consumption of all raw materials. The
United Kingdom has introduced already an "Aggregate Levy" of
€2.3 per ton for the domestic extraction of non metallic minerals for construction. A taxation of all raw materials including
minerals, ores and fossil fuels like coal, oil and gas would be
a consistent alternative to the existing environmental policy in
Europe, which is dominated by carbon emissions trading of the
primary industries in the ETS. This system will not be able to
reach the targets alone, so that a number of additional instruments will be introduced in the different countries. Further the
ETS is an instrument with very high administrative costs: more
than 12,000 installations which burn fossil fuels have to be under
control. A resource tax would be much easier to handle, because
the extractions and imports of resources would affect a reasona-
ble number of firms. And finally we should not forget that the big
problem of global warming is only a symptom of the process of

permanently rising resource extractions. Therefore our therapy should be targeted primarily to raise resource productivity to reach a decoupling of economic growth and resource extraction.

It is plausible that the damage to nature is strongly correlated with the weight of given materials, independent of their type. For all materials it is true that the extraction, transport and disposal have severe consequences for energy consumption, CO_2 emissions, dust, and noise and biodiversity. Therefore a material input tax has to be based on physical terms like tons. Taxing the extraction and imports of materials with a certain amount of euros per ton let the costs spread over all stages of production, so that prices of all products rise due to the direct and indirect materials which are part of the products. This will induce a reduction of material inputs including fossil fuels.

The implications for international competitiveness of finished goods are in general the same as in the case of the ETS regime.

In any case, there is something intriguing about the concept of taxing resource utilization while simultaneously easing the deployment of labor: by burdening companies in terms of one production factor, it is possible to steer towards our ecological goal, while at the same time reduced labor costs would be an offset for the company, and the labor factor would become more attractive. This would be a result of lowered social security rates, which in turn reduces the so-called non-wage labor costs. The subsequent gaps in the social security funds are then topped up by the revenue collected from the ecological tax. But there are some arguments against this way of refunding the tax revenue: first of all, reducing social security contributions and simultaneously topping up social insurance with tax funds would gradually introduce changes to the financing of the social insurance system. This method of ecological tax reform only makes sense if it is really desired. If this is the case, then the system should be changed independently of the eco-tax revenue. Secondly, unemployment is not a problem in some European countries, which is why it is not necessary to reduce non-wage labor costs.

The tax revenue could also be used to reduce income taxes. In this case it would not be concentrated on labor income, but would compensate negative effects of taxation on income generation in general. It can be expected that the higher net income will reduce the push in wage bargaining and will in this way reduce labor costs for firms.

Another form of using the tax revenue is the reduction of value added taxes. The taxation will raise prices depending on the direct and indirect amount of the use of nature that is incorporated in the product in question. A reduction of value added taxes with the revenue of the eco-tax would lower consumer prices so that there would be no increase of the general price level, but only the wanted change in relative prices.

If the target of refunding tax is to compensate the negative impacts on the taxed industry directly, then the refunds should be paid directly to the companies in the affected industries, as we have just suggested in the case of revenues collected by the state from auctioning off certificates. A company's turnover or output could be the key used to gauge repayments from the tax revenue. As a result, only the companies with ecologically "bad" technologies would be penalized, while the remaining companies within the industry would profit, and the industrial sector as a whole would remain unencumbered.

State-operated efficiency agencies

The wide consensus in literature is that considerable amounts of both materials and energy are being wasted. Unnecessarily heavy machines and vehicles are the cause, as well as wasted materials by the manufacturers during the production process, for example, in the construction industry, and insufficient recycling of used materials.

Estimates have been made as to the extent of this inefficiency, based on the experiences of renowned consulting agencies: the business consultant Hartmut Fischer, and other representatives of the profession, have identified a savings potential of 20% of total material costs in the manufacturing sector. Permanent cost reductions could be achieved with the help of extra consultancy services and investing capital amounting to the savings that could be made in a year. A third of these investment costs would cover the consultancy services, two-thirds would be additional capital costs.

One might ask why companies are nowhere near close to their optimum. The answer lies in the inefficient incentive mechanisms concerning material utilization within management systems. Generally, controlling systems do not explicitly focus on material losses, they focus mainly on labor factors and the reduction of labor costs. The reason behind this is that in the past labor costs have constantly risen, and while the price of raw materials has admittedly seen some quite heavy fluctuations, there have not been any noticeable upward trends up to now. Also, the decisions made over which machines to invest in are often governed by the acquisition costs, whereby the total operational costs covering the equipment's entire lifespan are often not paid adequate attention. Frequently management is not aware of all other technical alternatives and the implicated costs. Occasionally there are not enough institutional establishments for the exchange of information, which can particularly affect smaller businesses.

Seen from this perspective the markets are obviously not in the position to achieve optimal resource utilization. Taxing material utilization, as we have just discussed, would give added incentive to overcome these shortcomings. It might also make sense for the state to facilitate an information program that highlights the importance of material management and indicate which course would deliver improvements. Obviously, companies will have to cover the consultancy costs.

What kind of results could be achieved with such a program, assuming that the consultancy companies' experiences would be applied to all firms in the manufacturing sector? Martin Distelkamp, Marc Ingo Wolter, and myself addressed this issue for the German economy, and – casting a glance at the sponsor of the study (Aachen Foundation Kathy Beys) – called it the "Aachen Scenario". In order to evaluate such impacts, a model is needed that can represent an economy, and is able to describe how the economy functions as realistically as possible, in as much detail as is necessary. The PANTA RHEI system, developed by the Institute of Economic Structures Research (GWS) in Osnabrück, is one such model. It describes economic developments with a sectoral disaggregation of fifty-nine industries, their technological links to one another, their consumption of raw materials and energy, and connections to macro-economic developments. Moreover, the behavior of private households and also of the state is illustrated in great detail. The model's name is not, as is often the case, an abbreviation, it refers to a quote of the Greek philosopher Heraclitus (ca. 535–475 BC), which in English means "All things flow." This name was chosen to reflect the model's ability to identify changes in production and demand structures, and their interconnections with the environment.

Using this model, first of all a forecast was drawn up of developments up to 2020 without taking this information program into consideration. A second calculation was compiled including this information program, whereby, according to the experience made by the consultancy companies already mentioned, the following assumptions were developed: in the forecasts made a few years ago, it was assumed that each year, from 2005 onwards, approximately 9% of companies in the manufacturing trade would participate in the information and consultancy program, so that by the year 2015, eleven years later, the program will have reached its conclusion. In the first year, the companies in the manufacturing trade would have higher capital costs for the deployment of new machines, which would amount to just about 20% of the savings made in material costs. In the following years the reduction in material costs continues, whereas the additional costs for consultation and capital expenditure no longer apply. In the case of energy costs, consultation and capital outlays are higher and are equivalent to the material savings made over six years.

A comparison of the first forecast with the second then shows all of the direct and indirect effects as a result of the information and consultancy program, since this is the only difference between the specifications for the two computations.

The measures directly affect macro-economic development in two ways. First of all it reduces costs in construction, public administration, and the manufacturing trades. Secondly the turnover in the industries producing material goods is lower. Consequently, there are both winners and losers. But the winners are exclusively national companies, while the losers are both domestic and international companies. Consequently the direct effect would bring about an increase in the gross domestic product.

There are a lot of indirect effects involved with programs designed to dematerialize production. Figure 5 presents a graphical depiction of the most relevant interrelations. Initially, lowering average costs reduces prices. Should prices fall less than the costs, which is usually the case, company profits increase. This leads to an increase in income for households, who are the

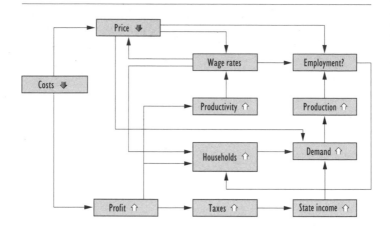

Source: Meyer, B., Distelkamp, M., Wolter, M. I. (2007)

Figure 5   Indirect macro-economic impacts of dematerialization

recipients of the profits, and also an increase in tax revenues for
the state. Both effects regenerate production, employment, and
demand for goods. Higher profits lead to higher value added and
increased labor productivity, which means an increase in value
per man-hour. These factors, and also price development, are the
most important influencing variables in wage negotiations. The
decline in prices also lowers wage rates, but increased labor productivity has a positive effect during the wage rate negotiations.
Both effects counterbalance each other, which leaves the wage
rate relatively unchanged. Consequently, the ratio between wage
rate and price level increases when prices fall. This is also called
the "real wage rate," because it tell us how many units of goods a
worker can buy for an hours work, and how many units of goods
a company needs to produce for an hours work. When the real

wage rate rises, the production factor labor becomes more expensive and there is less demand from companies. However, because production increased in our analysis at a much higher rate than the real wage, there was a rise in net employment. Increased employment raises household income, and consequently, once again, the demand for goods.

The program's economic impact is clearly positive: during the duration of the program, the growth rate of gross domestic product was each year approximately 1 point higher than in base prognosis. So instead of having an average growth rate of 1.7%, it would then be 2.7% per year. The rate of employment continues to rise, and by the last year of the program there are around one million more people employed than in the base prognosis.

The ecological results are also very positive. Figure 6 shows the development of total resource utilization measured in tons, including the resources indirectly contained in imported goods. The development of the base course demonstrates that, without the Aachen Scenario, resource utilization continues to increase. The use of metals plays a big part in this development, which is primarily due to increased demand from abroad for the production of capital goods. Starting from 2005, the development of the Aachen Scenario clearly diverges from the base prognosis. There is a drop in resource utilization, which only starts rising again after the program's completion in 2016.

The German Council for Sustainability has announced a target for improving material efficiency – i.e., the ratio between gross domestic product and total materials used – derived by doubling the 1994 figures by the year 2020. Figure 7 demonstrates this development, along with the PANTA RHEI forecast for material productivity as the base course, and the Aachen Scenario. We can see that without the measures a wide gap spreads between the target line and the base course. Yet even the Aachen Scenario

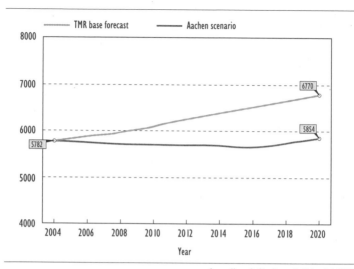

Source: Meyer, B., Distelkamp, M., Wolter, M. I. (2007)

Figure 6   Prognosis of the development of Germany's resource
utilization up to the year 2020, in millions of tons

is still a long way off from reaching the set targets. In this respect
it is clear that the information and consultancy program needs
supplementing with further measures.

The consultancy and information program is initially only
concerned with using currently familiar technologies as efficiently as possible. As we have seen, this is nowhere near enough
to reach the targets set by the Council for Sustainability, let alone
those defined by Friedrich Schmidt-Bleek. This is why we need to
speed up innovation. Here too, efficiency agencies could act as
facilitators, by assisting exchange between scientists and companies to solve technical problems. The effects of innovative technical changes are, from a qualitative point of view, equal to those
we have just discussed. Yet they could be considerably better.

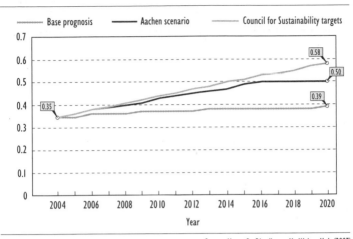

Source: Meyer, B., Distelkamp, M., Wolter, M. I. (2007)

Figure 7   Council for Sustainability targets and prognosis of material productivity in Germany up to the year 2020 in euro per kg in constant prices

Subsidies for the use of innovative technologies

Subsidies are benefits from the state for companies, with no service expected in return. Economists are no fans of subsidies, since they alter the price system and consequently create production structures that deviate from market-created efficient structures. If the only object of distributing subsidies is to secure jobs, then one really does need to take a critical stance. Subsidies are financed by taxes, which ultimately come from economic sectors more efficient than the sectors receiving the subsidies to secure jobs. This in turn puts jobs in other areas at risk, although these job cuts would go unnoticed, because they are spread across the entire national economy.

Therefore the distribution of subsidies can only be justified through further argumentation, which sometimes proves to be somewhat flimsy. One might question whether the long-term subsidization of jobs in, say, German coal mines can be sufficiently justified by arguing that it is in the name of energy security. The promotion of renewable energies is a different matter altogether. It is obvious that generating electricity from wind power, solar plants, photovoltaics, geothermics, and biomass (for descriptions of these technologies, refer to Chapter 5) is not yet competitively viable in comparison to conventionally produced electricity. On the other hand, it is to be expected that the price of fossil fuels will continue to rise due to increasing shortages and, above all, as a result of emissions trading. Technical developments in renewable energies, and the efficiency gains attainable through increased production numbers, will lower the costs of renewable energies, reversing profitability in the long run. Moreover, politics has already committed itself to raising the amount of renewable energies used for energy supplies in Europe to 20% by the year 2020. If this goal is to be implemented, then obviously the necessary measures have to be made today.

In Germany, for example, renewable energies comprise just under 5% (2005) of overall primary energy consumption, and 10.4% of electricity consumption, which makes it the world's leader in renewable energies. This position is mainly the result of the Renewable Energy Sources Act (EEG) of 2000. The act guarantees that operators of power plants with wind turbines, solar radiation, biomass, geothermal, and methane gas have long-term purchase prices to cover the costs. The next closest operator of a suitable grid is responsible for the acceptance and remuneration of the electricity created. The same applies to the next closest transmission network provider of high-voltage power supplies. The balance between the compensation rate and the market price of electricity is equally divided between the power supply companies and added on to the price of electricity. Accordingly, the consumers of electricity carry the costs. In this sense, these are not strictly genuine subsidies, because instead of carrying the costs, the state merely guarantees that the extra costs above the market price are covered by others. Financial assistance currently amounts to around 2.6 billion euros.

The financial assistance from the EEG should be retained if the patronage of renewable energies within the framework of emissions trading does not come into fruition as we have suggested. The EEG has already enabled innovations today in a sector that might otherwise have not been competitively viable for another ten years. The risks usually involved with the promotion of innovation no longer apply here, because the expected long-term increase in fossil fuel prices will sooner or later aid profitability, and it is also expected that further cost-cutting improvements will be made in plant technologies.

For many years the renovation of housing facilities has been an important element in the entire structure of climate protection policy in view of CO_2 emissions. After all, this constitutes approximately a third of Germany's total final energy consumption. Energy costs and CO_2 emissions can be cut by up to 50% with the help of innovative building insulation, triple glazing, highly efficient heating systems, avoiding thermal bridges, and with the use of ventilation technology with heat recovery. Funding is aimed at those home owners who were able in the past to apply for subsidized interest from the German reconstruction loan corporation KfW (*Kreditanstalt für Wiederaufbau*); the current program allocates redemption allowances. The program's format is subject to a certain amount of change. In 2006, funds of 1.5 billion euros were available. Without a doubt, the field of subsidizing innovative behavior modification should be continued and further expanded.

Research funding

Why do governments sponsor company research? Isn't this kind of subsidization for companies problematic, because in the case that the research proves successful, the sponsored company is the one that reaps the profits? On the one hand, state commitment in an area that is actually the responsibility of companies can be justified by the positive external impact of innovation. This means that the successful introduction of new products and production methods does not just benefit the companies responsible, the benefits also spread to other companies. Increased employee knowledge, better quality products and machines, which in turn are applied in other companies, all go towards improving the competitiveness of an entire economy. On the other hand, some intended developments often do not come to fruition since the success of internal research is often uncertain, or alternatively it is unlikely that they can be made within a company's planning horizon.

In Lisbon, 2000, the heads of state and government in the European Council decided to make the EU the most competitive, knowledge-based region by 2010. The seventh support program, starting in 2007 and extending to 2013, was set up to reach this target, and has a budget of fifty billion euros. This means an increase in funds of 40% compared to the sixth support program. Under the heading "working together," thirty-two billion euros have been made available to help boost research cooperation, especially for small and middle-sized companies. The central focus is on the previously mentioned key information and communication technologies, nanotechnology, energy, but also the health and traffic sectors will be given particular attention. A further 12.1 billion euros have been reserved to fund basic research. Some 4.2 billion euros have been set aside to boost the innovation potential of research facilities.

To continue with the example of Germany, funding for innovation has traditionally been the duty of the Federal Ministry of Education and Research. Alongside basic research, its conversion into market compatible products and technologies is also included for funding. With regard to our context, we shall only be focusing on the latter.

In 2006 the Federal Government decided to set up a program to sponsor innovation in the key technologies (whose advantages we already discussed in Chapter 5) in a specific manner between 2006 and 2009, with funds totaling 14.6 billion euros. These funds are earmarked for nanotechnology, biotechnology, as well as information and communication technology. The fourth key technology, discussed in section five – renewable energy – is already receiving funding from the Renewable Energy Act. The program is especially aimed at the use of key technologies in various industries. Particular mention is given to the export-intensive sectors of auto and machine manufacturing, as well as energy and environmental technologies.

The most important instruments are the specified programs research and development. Here the field of research is divided up into many separate topics, and grouped into research programs based on research hypotheses bringing together representatives of the sciences, business, and politics. Individual projects are then announced for promotion. There is a strong wish for science and business to cooperate in joint projects. In this respect Germany is lagging behind in the so-called "technology transfer" from universities into businesses. At the same time, there are hopes that small and medium-sized enterprises (SME) will receive funding through such projects. This would be profitable for both parties. Scientists are directly confronted with problematic situations in business, and businesses have direct access to elite research. Because the projects are assessed by independent scholars before being awarded funding, and the projects are carried out under the scrutiny of experts, the chance of funds being misused has virtually been eliminated.

Of course, there is not enough space here to look in detail at the entire list of the research support measures from the other ministries. An undoubtedly important program covers "hydrogen and fuel cell technology." In ten years a total of one billion euros is expected to have been invested, half of the funds coming from the Federal Government and the other half from twenty-four companies.

The certification of consumer goods, durable goods, and buildings

The economic instruments to increase resource productivity we have mentioned so far will only be effective if consumers, manufacturers, and investors are aware of the ecological consequences of their alternative options. Above all, private households, but also companies, often have little access to the information they need in order to identify resource saving alternatives when deciding which consumer goods, vehicles, household goods, machines, and buildings to purchase. In this respect, although political measures to heighten awareness might appear at a first glance to be the "softest" economic instruments, when combined with other instruments they may be the most effective.

So-called product-related environmental information is on the one hand an important instrument, helping businesses to build customer relations with the aid of a positive certification labeled on their products, and also helps strengthen a company's credibility. On the other hand, consumers rest assured that the product can stand up to claims made on its label, since internationally recognized standards are the basis of certifications, which by the way adhere to the regulations of the German Unfair Competition Act. Consumers buying a refrigerator in Germany today can be certain they will be fully informed, when buying in a store, of any ecological consequences they might want to take into consideration. It lies in our own hands how much important information we wish to receive.

The voluntary self-commitment of businesses to the DIN EN ISO 14020 standard is a guarantee that the following nine principles are abided by:

1. Claims made about a product's environmental properties have to be exact, correct, verifiable, and not misleading.
2. Trade barriers may not be created by the allocation of an environmental label.
3. Statements concerning the environmental properties of a product must be based on generally accepted, scientifically verifiable methods. The methods used, and the evaluation criteria have to be accessible.
4. To determine the environmental properties of a product, the product's entire biography must be supplied.
5. When possible, carbon footprints should be used in the analysis.
6. Environmental labels should not pose a restraint on a product's further development.
7. Administration overheads during the certification procedure should be kept to a minimum.
8. The entire contract awarding procedure should be transparent.
9. All of the important information concerning the respective environmental claims of a product have to be made available for potential customers.

The German Federal Ministry for the Environment (*BMU*), the Federal Environmental Agency (*UBA*), and the Federation of German Industries (*BDI*) have published a brochure, aimed at companies, which differentiates between five types of product-related environmental information. Environmental labeling according to Type II (DIN EN ISO 14024) is mostly aimed at the end user and focused on environmental aspects. The specifications are the sole responsibility of the producers. Environmental labeling according to Type I (DIN EN ISO 14024) is aimed at private and commercial end users, featuring one or more environmental attributes, and allocated by independent organizations. Declarations according to Type III are intended for commerce, trade and consumers. They encompass extensive information based on carbon footprints, whereby valuations are omitted. The specifications are the responsibility of the producers. Carbon footprints (DIN EN ISO 14040–14043) have been developed for scientific, economic, and political experts. The specifications are the sole responsibility of the contractors, but have to be inspected by independent third parties. Information from environmental management systems (DIN EN IOS 14001 and EU Eco-Audit) are intended for developers, purchasers, marketing experts, and administrative bodies. The producers provide the specifications, which then have to be inspected by environmental evaluators.

Examples of the environmental label type I are Germany's *Blue Angel*, Scandinavia's *The Swan*, and the EU Eco-label. The *Blue Angel* is the world's oldest and most successful environmental label. By the year 2007, 3,364 products from 126 commodity groups had been certified, ranging from paints and varnishes, through PCs and printers, right up to heating systems and ocean-going vessels. When the *Blue Angel* was founded in 1977, Germany did not even have a ministry for the environment, which meant that at that time its initiator was the Federal Minister of the Interior. Allocations are made by an independent jury, consisting of representatives from environmental and consumer associations, trade unions, industries, trade, business, communities, media, the church, and two federal states. The Federal Environmental Agency organizes the functional preparations in view of the allocation criteria, and the contractual processing of the use of the seal lies with the RAL German Institute for Quality Assurance and Certification.

A main feature in buildings is energy efficiency. This is why, since 1995, it is compulsory for new buildings to obtain an energy pass. In the course of implementing the European Directive 2002/91/EG, plans are also expected for an energy pass for building assets up for sale or rent. Firstly, documents certifying a building's energy efficiency would be an exceptionally important asset for potential leaseholders or buyers. Furthermore, this will give added incentives to building owners and sellers to invest in energy saving measures.

The selection of technical standards for vehicles, buildings, and equipment

The technical features of our vehicles, buildings, and durable consumer goods such as washing machines, refrigerators, and stoves, play a considerable part in the consumption of resources. With the use of economic instruments, the state can directly influence the extent to which these facilities are used. For example, the fuel taxes influence the distances traveled, possibly even driving styles too – but initially they will not affect the technical features of the vehicle. More expensive fuels will only play a role in a vehicle's technical equipment when it needs replacing. In this context, consumers can opt for more efficient vehicles, and, combined with mobility behavior, can reduce their resource consumption. In the past we have seen that this was only the case to a restricted extent. What were the causes? There definitely was not the kind of reorientation whereby owners of large executive vehicles replaced them with mid-range vehicles, and owners of mid-range vehicles settled with compact cars. Is this because there was a lack of more efficient cars in the desired category, or did consumers continue to place high value on performance? This is the core point in discussions over the failure of the automobile industry to meet target agreements on fleet consumption. Occasionally mention is made in the debate of setting a binding performance target for fleet consumption. The automobile industry is accused of not having steered technical innovation towards resource conservation. The automobile industry defends itself by arguing that carmakers can, after all, only produce the kinds of automobiles that consumers demand.

This example demonstrates clearly the problems associated with regulatory law. It makes sense for an automobile company to offer a range of different kinds of vehicles. Even if they lower fuel consumption in each class of vehicles, the fleet consumption can still increase if there is a rise in consumer demand for that particular sector of high-performance vehicles. One should be able to rely on a consistently continued policy of shortening fuel to prompt car manufacturers to pick up the right signals for their product development. The introduction of CO_2 emissions trading for private households would be the ultimate cue that might have the desired impact. Moreover, tough international competition between vehicle manufacturers will set the right course.

These statements not only apply to the automobile market, but ultimately for all facilities and systems where performance plays an important role in resource consumption. Yet again, the importance of strong economic regulations should be reiterated.

However, the rigorous application of these measures requires courage and assertion. If this should become a reason to discontinue further development, then regulatory law will have to ensure relevant product developments are made in order to meet targets. In the sector of energy-run products, the EU has already developed the so-called Eco-design Directive (Directive 2005/32/EG of the European Parliament and Council from July 6 2005). The directive's implementation regulations lay down which characteristics are responsible for a product's energy efficiency and other environmental factors. Implementation measures are already in place for kettles, boilers, hot-water boilers, PCs, monitors, televisions, battery chargers, office lighting, street lights, air-conditioning, electric engines, industrial refrigerators and freezers, household dishwashers and washing machines.

There are plans to copy Japan's successfully introduced "Top Runner" model: the most efficient product on the market is declared the standard which has to be met by all of the other products in the same category of goods, within a set deadline, e.g., five years. The products are very clearly defined, as for example with washing machines, where differentiations are made between machines that hold 4kg, 5kg, or 6kg loads. It is possible that using such a specific approach when setting technical norms for pollutant emissions in automobiles would make sense, and lead to a more detailed classification of vehicles.

The Top Runner Program encourages technical developments by setting companies exacting, yet attainable targets, while these targets undergo further development in the course of the competitive process. The advantage, compared to emissions trading and taxation, is the avoidance of distorted international competition because the same terms apply to both domestic companies and importers. Japan has achieved an increase in the energy efficiency of air-conditioners by 63%, and 83% with computers.

Education for sustainable development

Up until now we have contemplated how we can restructure our economy to optimize resource conservation by giving incentives (economic instruments) and making constraints (regulatory law). The most ethical path is obviously the one that convinces us to change our behavior. It is all about the intrinsic motivation to increase resource productivity arising from the understanding of its necessity. Obviously, it is no small task to persuade a large segment of the population. On the other hand it would offer immense leverage, because if consumers really want something, companies are forced to follow.

Naturally, a more sustainable lifestyle has to be learned. Protecting the environment needs to find a fitting place in our canon of values, but we also need to know a lot about the links between technologies, social impact, and biology if we wish to devote ourselves to sustainable lifestyles. This makes it clear that we will have to confront the topic at a young age, as early as childhood – ideally as early as kindergarten – and continue through all levels of education up to, and including adult education. Already in 1992, at the Rio Conference, the Agenda 21 called for a realignment of education directed towards sustainable development. The United Nations have declared the years 2005 to 2014 as the global decade for *Education for Sustainable Development*. Its implementation in Germany is run by a national committee comprising representatives of the Federal ministries, the Bundestag, the Federal States, non-governmental organizations, the media, the economy, and the sciences. The plan of action drawn up by the national committee has set strategic targets for the decade, and which measures should be taken.

Among others, the targets include: from the nursery schools upwards, the topic of sustainability should be taught in all schools. Furthermore it should become an integral part of teacher training. In the process efforts will be made to make use of existing local Agenda 21 networks, involving the local communities, associations, federations, and cultural establishments. Partnerships with businesses are additionally planned. International cooperation activities are also to be strengthened.

Sustainability and business management

Companies need to operate within their sales and sourcing markets, and in the process have to maximize their profits according to economic performance demands within the framework of the existing legal system. This leads to the definitive issue of maturity, of the time frame in which profits have to be maximized. Occasionally, the following calibration is made: the maximization of short-term profit serves the interests of financiers, which is sometimes called the maximization of the share holder value, whereas the long-term orientation of companies favors the interests of the stake holders. Stakeholders are people, or groups, that have claims concerning a company's employees, the state, or even society as a whole. Obviously sustainability is a long-term concept, which is why from this perspective there might be a conflict between being oriented towards the interests of the equity holder, and following a sustainable company policy.

One might doubt whether this calibration pays off. First of all there is no reason why following the financier's interests leads to a short-term oriented business enterprise. The deciding factor for this classification is whether the financiers are looking for a long-term successful investment, or are only interested in maximum profit within a short period of time. Are the owners institutional investors, controlled by fund managers, whose success depends yearly on the market value of the funds? Or are the financiers individuals and families looking for a long-term investment of their capital? Ultimately the decisive factor for the fund manager is the value of the company on the stock market, which is naturally governed by the short-term annual, and even quarterly, success figures. The concept of sustainability only has a place in this environment when it is imposed by a legal framework. In the case of more medium-sized structured ownership, long-term oriented investment offers enough leeway for the provision of a sustainability concept. Efforts to strengthen equity holders' intrinsic motivation for a more sustainable company policy can be more successful under such conditions. The responsibility towards the company and its employees is certainly more pronounced with this structure.

One should also consider that Article 14, paragraph 2 of the Federal Republic of Germany's Basic Law stipulates that: "Property imposes duties. Its use should also serve the public weal." Obviously this also applies to the publicly quoted stock corporations governed by funds. In this respect, it is time to reconsider expanding this constitution article towards sustainability targets, and to bring it into the extensive debate over corporate governance. Corporate governance means all of the values which constitute responsible business management. In the future it will all come down to renowned entrepreneurs and business men setting an example by publicly extolling and implementing these views.

7 A Country in Focus: Germany
What Changes Will Have to be Made to the Labor Market and Social Security System?

The labor market and demographic change – a status quo forecast

We have advocated a dynamic strategy to reorganize the world economy, with Europe as the driving force behind it. We have given detailed explanations of the measures which should be taken to improve the willingness of companies to make innovative changes. In the process it became clear that this concept can only work within the framework of unrestricted international competition. As we have already seen in the first chapter, technological change always brings about changes to social structures, which always causes insecurity. For example, Germany's population is already uneasy in view of the risks posed by international location competition. However, the success of Germany's exports continues to offer new opportunities to flexible and highly qualified employees. The problem lies in the fact that although our strategy offers better opportunities for the upwardly mobile, it also increases the risks for the underskilled.

How much will we have to adapt to meet the expected qualifications demanded in job specifications in the coming decades? In this context, how much of this will be the result of demographic change, and what role will economic changes play? Together with Marc Ingo Wolter, I set about tackling these questions in the

Age in years	December 31 of each year					
	2001	2010	2020	2030	2040	2050
Younger than 20	17,259	15,524	14,552	13,927	12,874	12,094
20–35	15,925	15,445	14,860	13,254	12,639	12,086
35–50	19,647	19,060	15,691	16,064	14,569	13,574
50–65	15,543	16,448	19,500	16,361	15,672	15,123
20–65 total	51,115	50,953	50,051	45,678	42,880	40,783
65 and older	14,066	16,589	18,219	21,615	22,786	22,240
Total extractions	82,440	83,066	82,822	81,220	78,539	75,117

Table 10 German Federal Statistical Office's 10th coordinated population projections, according to age groups

Source: Federal Statistical Office, Germany (2002)

German context. Initially we were only concerned with changes expected to the status quo, in other words, we did not take into consideration programs designed to increase productivity or any changes to the education system.

Table 10 shows the German Federal Statistical Office's tenth coordinated population projections up to 2050, divided into different age groups. This projection uses mean assumptions of life expectancy and the annual immigration balance (+200,000). By the year 2050 the entire population will have shrunk to 75.1 million, while in the age group of over sixty-fives there will be an increase from fourteen million in 1991 to 22.2 million. At the same time, the age group twenty to sixty-five, which generally covers the working population, will fall from 62% in 1991 to 54% in 2050.

What effects will these changes have on job prospects if we take today's preexisting education system and assume that jobs will be made readily available? We chose the year 2000 as our point of reference due to the availability of all of the necessary data. We used the population model DEMOS to generate our forecasts based on the structures from 2000. Furthermore, we asked which developments in labor demands are to be expected with a continuation of the current economic framework. We made a "business as usual" projection by way of the previously mentioned environmental and economic model PANTA RHEI, which is able to predict economic structural changes. This gives us an estimate of the effect demographic and economic changes will have on the future labor market without including any innovation strategies. Based on these results we can then tell whether the requirements for the innovation strategy can be met by the labor market.

Let us start first with the developments in labor supply in the status quo forecast: it is assumed that the structure of professional graduations in the different ages of the population per year will stay constant. It is also assumed that the structure of labor market supply, according to age and gender, will remain unchanged. For example, we can assume that the percentage of women aged twenty-five offering their skills to the labor market in 2030 will be the same as the figures in 2000. It can also be assumed that the number of hours a twenty-five year old with specific qualifications works per year will be the same as those in 2000. Differences in labor supply, measured in hours per year, for people with varying qualifications will then only arise as a result of the population figures in each age group, according to demographic change. By using this approach we are able to answer the question of the extent to which demographic change will affect labor supply, depending on the age, gender, qualifications and amount of hours a person works.

In the study, qualifications are divided into an international scheme of six levels. We will slim these down into three groups: low, middle, and high qualifications. The maximum level of

education of those in the lowest group would be junior high
school completion or a high school diploma, without any additional vocational education. The middle group contains either
an advanced technical college certificate or higher education
entrance qualification with or without vocational training, a
completed apprenticeship, a qualification for a profession from
a vocational school or a one year course at a public health care
school. Qualifications in the top group include are master crafts-
men or technician education, specialized vocational college
certificate, completion of a two or three year course at a public
health care school, graduation from a technical college or a
university of cooperative education, or a technical college or a
college degree.

First of all, it became evident that the number of people in
the lowest qualification group will decrease by approximately 8.2
million between 2000 and 2030, whereas the number of people
in the middle group will increase by approximately four million
and the highly qualified will go up by approximately 2.5 million.
The reason for this lies in the high proportion of poorly qualified
people – especially women – in the older age groups who will no
longer be alive in 2030. Furthermore, it became noticeable that
the younger generations tend to be better educated.

It is a completely different picture with regard to the qualification levels of the working population due to differences in age,
gender and qualifications. First of all it should be noted that the
total group of employees in 2030 will be 5.6 million less than in
2000. When compared to population developments, this heavy
decline is due to the fact that the fifteen to sixty-five year olds will
be particularly impacted by demographic change. This explains
why we can expect declines of 0.6 million in the highly qualified
working population, and 2.2 million in the middle group. Sur-
prisingly, the group of low skilled employees will only decline

by 2.6 million, while the total population figures for this group are expected to decline by 8.2 million. There are two reasons for this: the upper ages in the group of under sixty-fives have low employment levels. Plus, those in their thirties today have a relatively high level of education, and by 2030 they will then be in the upper age group of under sixty-fives.

Let us now turn to the projections of developments in labor demand. The PANTA RHEI model analyzes Germany's economy in a sectoral disaggregation of fifty-nine branches. The interrelations between each sector, the global economy, and the economic behaviors of private households and the state, are all reflected in detail. The determinants of labor demands in all sectors, such as wage rates and developments in productivity have been integrated into the analysis in all variables. As for the employees' levels of qualifications, we took the configurations from 2000 for each sector.

The middle group of qualifications was predominant in all sectors, but in the service sector the number of highly qualified was noticeably higher than in the manufacturing sector, whereby the number of moderately qualified and also lower skilled workers in the service sector are lower than in the manufacturing industry. Because the importance of the service sector will continue to increase in the future – its share in the labor market will increase from around 75% in 2000 to 82% in 2030 – so will the demand for highly qualified workers.

Let us now compare the developments in labor supply and labor demand. Because men and women with differing qualifications and of different ages offer different yearly working hours, we calculated both labor supply and demand into hours, and compared the respective values according to qualifications, without taking into consideration the implications of the differences in labor supply and demand. In reality this is obviously not

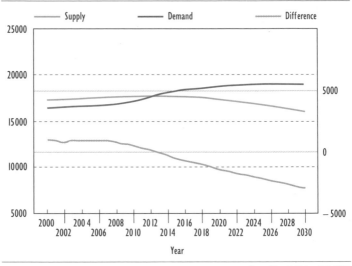

Source: Meyer, B., Wolter, M. I. (2007)

Figure 8   The development of supply and demand in Germany for the
highly qualified in millions of hours

the case, because wage rates and other economic factors change,
and consequently have an effect on supply and demand.

Within the projection's timeframe, the current considerable
surplus of supply will gradually shrink, because supply will
decrease and demand will stay relatively stabile. In 2030 only
around 600 million labor hours for the low skilled employees will
be unemployed, which works out at around 330,000 people if
we assume that someone with a full-time job works 1800 hours.
So the unemployment figures of what is currently a particularly
problematic segment of the labor market will improve decidedly.
The surplus offers for those in the middle group will stay the
same until 2010, but will then decline consistently, reaching a

comparatively low value in 2030 of approximately 1200 million hours, which in full-time equivalents would amount to around 660,000 people.

Let us take a closer look at the results in Figure 8 for the highly qualified group. The illustration has two different scales. The left-hand side of the graph shows the levels of supply and demand, the scale on the right-hand side shows the difference between the two. The fine line represents the developments in supply, the dotted line represents demand. The vertical columns show the yearly values in the differences between supply and demand.

Up until around the year 2012 there will be a small surplus of offers, which in the following years will rapidly decrease. A gap will appear and spread between the developments in demand and the supply of highly qualified employees. There will be an increase in demand for this group because of changes in the economy's structure, while demographic developments will lead to a decline in supply, however. By 2030 the demand gap will amount to around three billion labor hours. Converted into full-time equivalents, there will be a shortage of around 1.6 million highly-skilled employees.

Mobilizing labor supply and education campaigns

To recapitulate: the market for low and medium skilled labor will be substantially cleared, while in the sector for highly-skilled labor there will be a big unfilled gap in demand. In the process we viewed the normal economic developments that would take place without an innovation strategy designed to improve resource productivity. Even without a more detailed analysis, we can see that a systematic innovation strategy would increase the service sector contingency, especially in the areas of research and development, but also in general in the consultancy service companies working closely with businesses. Changing the structure of consumption in favor of services will increase the number of highly skilled persons employed. The demand for highly qualified individuals is likely to see a more dynamic development than in the status quo scenario.

Alongside the targeted innovation strategy to improve productivity, we will also need to increase total labor supplies. The willingness to enter into gainful employment has to increase, whereby we have to remember the present reserves of the female workforce, the reserves in weekly work hours and the number of lifetime work hours.

The number of women employed has scope for improvement, and needs considerable further development. In 2002, 73.6% of men and only 58.9% of women were employed in Germany. Setting up more daycare centers would not only increase birth rates, but would also increase the number of people able to work. Greater priority has to be given to the social responsibility of caring for children of all ages.

Shortening work hours, both weekly and lifetime work hours, has to become a thing of the past. We are already facing huge opposition to the recently extended age of retirement. Those who stand shortly before retirement feel justified in arguing that this is only an indirect way of cutting pensions, since nowadays most pensioners have stopped working before reaching the age of retirement. However, it should be noted that the majority of this generation has not managed to get to grips with the revolutionary introduction of modern information and communications technologies at the work place. The same cannot be said of today's thirty-year-olds. They expect, and are used to adapting to new software and hardware in the workplace every few years. It is to be expected that this generation will come very close to fulfilling the ideal promoted by education researchers of "lifetime learning." Naturally there are some professions where it is unrealistic, due to physical limitations, to expect that they will be pursued right up to the age of sixty-five, as is the case with roofers. Yet even a roofer can be retrained to enter a new profession at the age of fifty.

At the same time an educational campaign needs setting up to help meet the demands for highly skilled employees. If we can manage to do this, we will also be able to solve the problems in the area of pension funds, because demographic change will not affect the number of people employed too dramatically, and incomes (especially upper bracket incomes) will increase.

We have already mentioned increasing of the number of daycare centers, which is also relevant in this context. We cannot afford to keep ignoring the reserves of talent available in low-income backgrounds. This is why we need to introduce all-day school services to the general education system so that the children can receive comprehensive encouragement. The vocational schools and technical colleges will chiefly be concerned with offering a high standard of education, which calls for the provision of the necessary technical equipment, and that teachers receive regular further training.

Germany's universities have been undergoing dramatic changes over the last few years. The curricula are being adapted to integrate bachelor and master degrees, in order to facilitate easier academic exchanges for students and also for teachers, which is proving to be a strenuous task. But there is a comparative lack of funds for Germany to be able to match international competition. In the past policy makers have frequently cut university budgets to fill fiscal deficits.

Minimum income and the flexibilization of the labor market

We do not need to be worrying about the availability of work for the well educated. On the contrary: we have seen that there will be considerable excess demand on the labor market. But what will the situation be like for the poorly qualified? As we have already seen, the status quo labor market will be desolate. But if we choose an innovation strategy, not only will we need even more highly trained employees, we will also have less need for unskilled workers. So not only will there be a decline in the demand for unskilled workers due to demographic changes, it is possible that the innovation strategy might have an even more extreme impact on this sector. In this respect it is highly unlikely that this segment of the labor market will see a relaxation in the urgent issues of today.

If we want to keep this sector of the labor market flexible, then there will be a continued pressure on wage rates due to international competition. In some industries – e.g., in construction and industrial cleaning – this has already resulted in the introduction of a minimum wage rate. Today (spring 2008) there are heated debates over the general introduction of minimum wage rates. However, this would not be the right path. If the minimum wage rate is higher than the wages that would normally be generated by a totally flexible labor market, then there will obviously be less jobs for the underskilled than there would be if wages had developed freely in the labor market. This basic knowledge has continually manifested itself over the last decades during attempts to secure lower levels of income with disproportionate increases in wage rates. In fact, minimum wages bring about the exact opposite of its desired effect.

After all, the main concern is not how high wages are, it is more about how much income is available to those in the lowest income bracket. An alternative to the minimum wage is the introduction of taxes and social transfers within various models of the so-called "negative income tax." The basic idea behind both is: based on a minimum income unemployed persons receive, the poorly paid would receive a top up transfer that decreases with an increase in earnings. Ultimately a critical level of income is reached where the amount transferred, or respectively the negative income tax as the case may be, is zero. Once this point has been reached, the "normal" tax rate applies.

In fixing a minimum income rate, benefits such as nursing care insurance, health insurance, unemployment insurance, and pension insurance funds should not be taken into consideration because they are after all insurance benefits financed by contributions. The minimum income could be equal to the subsistence level, which would mean that at least the present welfare benefits would be integrated into the tax system. A negative income tax rate should be chosen in such a way as to give enough incentive to increase earnings, even if this means a reduction to the transfer. There are various suggestions published discussing negative income tax, the main difference being the amount of welfare benefits included. Obviously this would create considerable upheaval in the distribution of duties and responsibilities of the national government, federal states and municipalities in our federal polity, and which would also entail the need for reorganization.

Combination wages work in a similar way to negative income tax. Again the employee receives a subsidy from the state in addition to his or her wages. The only difference to the negative income tax is that this concept does not go as far as to include the entire social security system. It is already possible for recipients of *Arbeitslosengeld II* (a German flat rate unemployment benefit program) to earn a second income without restrictions on the amount of hours worked. In this way less cuts will be made to the unemployment benefits of those earning a second income than has previously been the case. It is hoped that this will help bring about a gradual reintroduction back into the labor market. This would mean that in the future countless employees could receive state subsidies. Therefore, the long-term unemployed can take up poorly paid work and receive extra subsidies for a maximum of twenty-four months, which is essentially the introduction of combined wages. As of spring 2008, negotiations are presently being made for the general introduction of combined wages.

Compared to the minimum wage, the advantages of combined wages and negative income tax are obvious: the labor market would set the wage rate that would provide full-time employment in the lower income bracket. At the same time, state subsidies would guarantee households an income above the level of subsistence, without dampening the motivation to find paid employment. We should bear in mind that these are only provisionary measures that will no longer be necessary in the long-term, since shortages on the labor market will lead to an increase in job opportunities with the implementation of the innovation strategy.

Problems for the social security system?

The social security system in Germany consists of nursing care insurance, health insurance, pension insurance funds, and unemployment insurance, and is a compulsory contribution for employees – although not all employees are subject to paying contributions. The self-employed, officials, and the marginally employed do not pay any contributions. There is an added allowance from tax revenues, which has developed over the course of time as a reaction to historical situations. Demographic change is placing a heavy strain on the system, because – as was seen in Table 10 – the number of people paying into the system will decline, while the number of people eligible for benefits will increase.

This will cause particular problems with the pension insurance funds. Politics has only just started responding over the last few years by making necessary adjustments, despite the fact that the problem has long been a well-known fact. There is a delay in the dynamics of pensions adjusting to general income developments, and the age of retirement is gradually being increased. At the same time the so-called *Riester-Rente* (Riester pension scheme) offers an attractive supplement to pensions, whereby the state makes an additional contribution to the savings of voluntary participants. Its name derives from Walter Riester, the former Federal Minister of Labor and Social Affairs, who originally suggested the additional pension allowance scheme. It was instigated by the Pension Reform Act 2000/2001, in which the pension of the average employee subject to, and paying, social insurance contributions for over forty-five years, will in future be reduced from 70% to 67%.

While the general public had completely ignored the problem for many years, there is now a tendency for over-exaggeration in public debates, damaging contributors' trust in the system's viability. This demonstrates the typical weaknesses of social criticism in German public dialogs: at long last we acknowledge a future problem, but discussions are essentially one-sided and completely ignore the interdependencies involved with other advancing developments, let alone consider possible solutions. Projections of future demographic changes speak for themselves and cannot be doubted. However, not only will age differences place a strain on the social security system, the amount of people employed and the level of their qualification will also play a decisive role. There are considerable reserves that we will have to mobilize, because – as we have already seen – highly-skilled work will become a much sought after production factor in Germany. From this point of view, if we can implement an innovative strategy to increase resource-efficiency, social security problems will take care of themselves. A course is starting to emerge, which could solve both ecological as well as economic problems, and many social problems.

8 Perspectives for More Sustainable Development in Europe

Estimating the potential of future developments using environmental economic models

We have demonstrated which instruments need to be employed in order to achieve a more sustainable development in Germany and Europe. But do these instruments really work, and on what scale should these measures be eventually combined? To determine the necessary scale and the exact form of the instruments requires very complex estimations. It is comparable to a doctor's medicinal treatment of a patient. The doctor has identified specific abnormalities in a patient's physical condition. The physician makes a detailed examination, until he or she finally pinpoints the cause of the ailments, and is then able to make a diagnosis. Based on this diagnosis the doctor then chooses a suitable method of treatment for his or her regimen. This is where we find ourselves in our pursuit of finding a regimen for our patient, the economy. We now have to apply the final stage. The dose of measures, their exact configuration, and the combination of measures all need taking into consideration, whereby – just as in the field of medicine – we have to evaluate the risks and side effects.

The doctor does exactly the same thing with the patient by taking allergy tests, or asking the patient of any past allergic reactions, in an attempt to restrict any possible risks. The physician can fall back on his or her own experiences, as well as that of colleges and the pharmaceutical company that made the medicine, in order to ascertain the compatibility and possible side effects that might arise from prescribing a combination of medicines. Our situation is much more complicated. The patient is silent, unable to take an allergy test, or similar tests, and very little former experience has been made to verify the compatibility of the instruments. This is the reason why politics sets targets much faster than it creates the specific measures needed in order to reach the targets.

Astute reflections alone are no great help here, because the circumstances are extremely complex. The economic impact of the various measures, and the influence they have on the use of natural resources are, in part, opposing. They are also partly complimentary, yet the results are often non-linear and often have a delayed impact. We presently find ourselves in the same situation as climatologists, who also only gain their knowledge based on model calculations.

In order to estimate the results of the implementation of the instruments, we need models that can portray real-case scenarios so that we can perform some experiments. Obviously these models would be simplified versions of reality, but it is decisive that the necessary facts are sufficiently represented, while less important facts can be neglected. A topographic map, for example, is a representation of the earth's surface, and in this respect a model of the Earth's surface. The differences between a hiking map and a street map of the same area lies in the fact that they were made with different applications in mind. Whoever sets off sailing on the North Sea with a road atlas is in for some nasty surprises, and would be better off using a nautical chart! Likewise, to navigate political policies we need a suitable map. Without one we will only get stranded.

What kind of demands will we have to make on models representing environmental and economic interrelations, allowing us to estimate the impact environmental policy measures will have on both the environment and the economy? If we wish to discuss environmental policy within the European framework, we will naturally need a representation of global environmental and economic interrelations, because Europe is a noteworthy part of the world, and its politics have an influence on the behavior of other countries. Furthermore, environmental problems are ultimately global problems, although it is at the same time essential for regional differences to be integrated into the analysis, because human behavior varies from region to region, and the adverse effects our behavior has on the environment varies regionally too. The following five criteria have to be met by a global environmental economic model:

1. The model has to give a detailed categorization of countries as well as regions, whereby it makes more sense to categorize according to countries since it is countries that are politically relevant units. Southeast Asia as a region is comparatively uninteresting, because it includes the countries of China, Japan, Korea, Thailand, Indonesia, Philippines, etc., which all have very different economic structures and environmental-economic interconnections, plus each country has different policies and goals, and employs different instruments.

2. The model has to represent an in-depth classification of a national economy's industrial and sectoral structure. Links between the economy and the environment, with the complex configurations between raw materials and emissions, can only be included once the main extractors of resources and emitters of pollutants have been identified, and the technical production links between the other industries have been mapped.

3. Global trade is the most important economic network, where the national economies of individual countries connect. We will also need an in-depth representation of the international trade of all countries, classified into groups of goods, to match our requirements.

4. The model needs to be able to explain economic developments and links between the environment and the economy. Many models make presumptions concerning economic growth on which they base their calculations, using preexisting industrial infrastructure. Obviously this is impermissible, especially if we want to find out the effects environmental policy measures would have on the environment and the economy.

5. The model has to be able to give a realistic representation of economic developments and the use of natural resources. It needs to be able to give a detailed enough explanation of observable historical developments.

The Japanese economist Kimio Uno has described the attributes of no less than thirty-four previously published global environmental economic models. If we take our list of requirements, only two of the thirty-four models remain, and consequently also have many similarities. The first of the two models is the GTAP system (Global Trade Analysis Project), which was originally developed by the Purdue University. The system was such an international success, that various versions are used today for diverse issues – including environmental economic issues using the GTAPE version. The second model is the COMPASS system (COMprehensive Policy ASSessment), or its successor GINFORS (Global INterindustry FORecasting System). COMPASS was developed by the GWS (Institute of Economic Structures Research) in Osnabrück, within the framework of a

Japanese/Chinese/Belgian/German cooperative project financed by the Japanese government. The follow-up model GINFORS was also developed by the GWS within the European MOSUS (MOdeling SUStainability for Europe) project.

The remaining thirty-one global models either do not distinguish in sufficient detail between countries, which are defined by regions or continents, or they do not include enough industrial sectors. They are often models that, although they include enough countries and industrial sectors, are unable to offer simultaneous explanations of economic and environmental developments. They are frequently models that give very detailed illustrations of energy usage during the production and consumption of various goods, and the consequent emission of pollutants, but are unable to provide economic explanations of the demand for goods and economic development. The latter is then furnished in some form by way of assumption.

Both the GTAP and GINFORS models, however, are based on two different philosophies. The GTAP model assumes that all manufacturers, consumers, and investors have the complete information necessary on alternative options, and are therefore able to make optimal decisions. All markets where the interests of suppliers and consumers are harmonized by prices are competitive markets, meaning that all market participants only have a minimal share of the market, and they are unable to noticeably influence the general market as a result of their decisions. Prices develop in such a manner that supply equals demand on every market. These, and a few further requirements, are the core of the so-called neo-classical modeling approach, which ultimately allows a company's behavior patterns to be derived from technological characteristics, and consumers' behavior patterns can be derived from their utility considerations. It is a closed modeling concept in which a few central conjectures ultimately define the model's structure. The equation's parameters are partly preset, the rest are designated to enable observations of the model's variables over the chosen year. Models of this kind are called computable general equilibrium models (CGE).

The model GINFORS is based on a different philosophy – that of limited rationality, which has its origins in evolutionary theory. Economic agents do not have full information about the alternatives they have. So it is not possible for them to realize the optimal decision, because the restrictions in information may exclude the best alternative and allow only "bounded rationality." This is the reason why economic agents tend to follow set routines when making decisions. This makes the modeling open, because it offers the creator of the model a wealth of more or less plausible hypotheses about the behavior of the agents. The "right one" can only be identified by conducting empirical tests. This means that the parameters for the behavioral equation are calculated on the basis of observations made over the longest period of time possible, using econometric-statistical procedures. Only those behavioral hypotheses which, compared to others, have been able to explain actual developments over a longer period of time are then used in the model. This is what makes GINFORS a so-called "econometric model." The prices in this model are not governed by supply and demand, but by a mark-up on the unit costs.

Both modeling versions are not only to be found in global economic-environmental models, they qualify for two general developments of environmental economic and also pure economic models. When discussing the performance of their models, advocates of the neo-classical modeling emphasize the closed aspect and the clarity, and consistency of their approach, advocates of evolutionary modeling place an emphasis on the empirical validity of their models. Both approaches are justifiable. Neo-classical models are recommendable if one wants to know the kind of effects a political measure would have under ideal conditions. The econometric models are more suitable for discussing the actual results that can be expected from measures, including any market imperfections. With evolutionary models, the question arises in projections and simulation calculations for the future of how long during the observation period the economic agent's behavioral patterns remain stable. In this respect the maximum period for future projections should not exceed twenty-five years. The neo-classical general equilibrium model can be used for longer periods of time, because the validity of a typical idealized behavioral pattern can be proved independently of time.

Neo-classical CGE- models as economic- environmental as well as pure economic models have been developed in great number all over the world. The evolutionary modeling philosophy is also pursued by the English institute Cambridge Econometrics, with the European economic environmental model E3ME, directed by Terry Barker. A pure economic system in this line is the global INFORUM International System, led by Clopper Almon from the University of Maryland (USA).

Now that we have taken a brief excursion through the world of modelers, let us take a closer look at the GINFORS global model, which we will subsequently be using to answer the questions we asked earlier. GINFORS depicts the economic developments of fifty countries, which include the EU-27, all of the countries in the OECD, China, India, all of the countries in Southeast Asia, Russia, Argentina, Brazil, Chile, South Africa, and the OPEC countries. The remaining global countries fall under the category "Rest of the World."

The economy has been divided into forty-one industries for twenty-four countries, which include all of the EU-15 countries (excluding Ireland and Luxembourg), all of the important

OECD countries, and China. World trade has been modeled bilaterally into categories of twenty-six industries. This enables the model to give an exact illustration, for example, of German car exports to the USA. The energy demands, national supplies, and the exports and imports of eleven energy sources are defined for each country. The model also calculates for each country the extractions of biomass, ore, non-metallic minerals, oil, coal, gas, and other remaining raw materials from the environment.

This model has already been used for many environmental economic analyses. In the following sections of this chapter we shall look at the results of simulation calculations for Europe within the framework of the MOSUS project, which were compiled using GINFORS.

The MOSUS Project – alternative scenarios for development in Europe

MOSUS is a project within the fifth framework program of the European Union, that has searched for European strategies capable of facilitating more sustainable development. The impact of various measures on resource utilization, pollutant emissions, and the economic and social developments of all of the EU-25 countries were investigated. GINFORS was the simulation model utilized, which we have just talked about. Twelve research institutes from eight European countries were involved with the project that was conducted from February 2003 through to January 2006.

The business-as-usual-scenario (BASE) portrays a world in which the current behavioral patterns of all economic agents, including politics, remain unchanged. The alternative scenarios outline future visions in which policies follow either relatively low, or high sustainability targets. The same political measures were taken for both scenarios, whereby only the dosage of the measures was varied. The various measures and autonomous developments which were observed can be summarized into six groups: technical change, transport costs, recycling and material efficiency, Aachen Scenario, research and development, and emissions trading.

The following developments were supposed in the sub-scenario "technical change," which was partially achieved with the help of research funding: the use of pesticides and other agricultural chemical products can be reduced by around 0.5% a year, thanks to continual biotechnological progress. Accomplishments in the development of incinerator technologies lead to a reduction in fossil fuel use by around 1% a year. Technical progresses in electric arc steelmaking will increase its share in the steel manufacturing industry, leading to an increase by 0.5% a year in the steel industry's demands on the recycling sector for scrap metal and electricity, causing a consequent decline in demand for coal and ore. The automobile industry will undergo substantial changes. New materials, such as polymers, will reduce the application of steel and other metals by 0.5% a year, whereas the use of electric motors, batteries, and electronics will increase by 1% a year. In "low" scenarios, these changes will start in 2015, and in "high" scenarios already by 2010.

The sub-scenario "transport costs" replaced the present taxes in the transport sector with a fee-per-kilometer. The revenue gathered by the state remained constant, but the price of transporting passengers and goods increased in the "low" scenario by 5%, and in the "high" scenario by 10% when compared with the base scenario.

In the sub-scenario "material efficiency," a tax would be levied on the utilization of metals and non-metallic minerals (sand, gravel, etc.). The sectors affected would be completely exonerated by cuts in other taxes, preventing the state from receiving additional tax revenues. This measure would improve metal recycling in the "low" scenario by 0.1% a year, and in the "high" scenario by 0.3% a year. The efficient use of non-metallic minerals would increase annually by 0.2% in the "low" scenario and 0.4% in the "high" scenario.

The "Aachen Scenario" project is an information and consultancy program moderated by the state for companies in the manufacturing industries. Annual material input is to be reduced as a result of consultancy. The costs arising from consultancy will increase the deliveries of the "research and development" services sector to the manufacturing industries, whereby the consultancy costs, which amount to a year's savings in material, are one-off, and the reduction in costs for materials used are naturally lasting. If an engineer improves a company's technology, he or she can only charge for these services once, yet the technological improvements obviously last longer. As a result, the company that received consultancy benefited from productivity gains. This implies that up until now companies have not yet fully realized their full potential in relation to the best technologies currently available. We have discussed this in detail in Chapter 6, page 97. In the "low" scenario the savings in material costs will be 10% by the year 2020, in the "high" scenario savings will amount to 20%.

In the sub-scenario "research and development", it is supposed that the European states will subsidize the research and development of companies by an additional 1% of state consumption, which would then be correspondingly lower. The hourly productivity of labor would increase by an annual 0.15%.

With regard to energy utilization and CO_2 emissions, it is supposed that emission trading will continue, whereby the price per ton of CO_2 emissions in the "high" scenario of 120 euros is obviously higher than that of 40 euros in the "low" scenario.

Furthermore it is supposed that the amount of biofuels in the "low" scenario equals 10%, and 18% in the "high" scenario.

The scenarios include a few of the measures we have already discussed in Chapter 6. Particularly the measures of research funding, information and consultancy programs have been considered here, but also emissions trading, which had just been introduced at the time that the simulation calculation project MOSUS was being designed. But the various regulatory measures, such as the specification of technical standards for household appliances, vehicles, and buildings, as well as promoting the intrinsic motivation of companies and consumers, have not been included.

The measures were distributed in a way to enable the "low" scenario to meet the Kyoto target for the EU-25, while the "high" scenario aims to meet targets set by the IPCC (Intergovernmental Panel on Climate Change) scientists in order to reduce climate gas emissions by 20% of the levels in 1990 in the year 2020.

Can Europe's targets for sustainable development be met?

All of the measures described are primarily aimed at increasing the efficient utilization of resources. Yet they have very different effects on economic development. The taxation of transportation and material input, as well as emissions trading within the manufacturing sector, all lead to more expensive energy costs, which in turn result in a rise in prices, and ultimately a reduction of resource utilization. This will also have a negative impact on the economy since higher prices curb production and employment figures. However, the impact would remain minimal, because the state keeps the level of its total revenue the same, by cutting other taxes in the transportation sector and the manufacturing industries, for example.

	Gross domestic product	Resource extraction	CO_2 emissions
Germany	5.1	−22.4	−19.2
France	7.8	−4.6	−18.4
Great Britain	1.1	−8.4	−12.6
Spain	6.1	−3.2	−15.4

Table 11 Effects of the "high" scenario of the MOSUS project on the gross domestic product, commodity prices, and employment in selected EU countries

Divergence from the base scenario of 2020

Source: Giljum, St., Behrens, A., Hunterberg, F., Lutz, C., Meyer, B. (2008)

Measures designed to support technical improvements help reduce production costs and in the process the price of goods, which always has a stimulating effect on economic development. On the other hand, the state has to make cuts in other areas in order to be able to fund research, which in itself has a negative impact on the demand for goods. In most countries the positive effects on economic developments prevail.

The Aachen Scenario information and communications program, designed to increase resource productivity, cuts the prices and raises the real net output, thereby generating decidedly strong and positive economic results. Consequently, economic and ecological impacts are coordinated and aligned according to this scenario. An increase in income and production slightly lowers the favorable ecological results, however. This is also referred to as the rebound effect.

Table 11 shows the effect the "high" scenario would have on the gross domestic product, resource extraction, and CO_2 emissions of the larger European countries of Germany, France, Great Britain, and Spain. The scenario would generate positive effects on the gross domestic product of all countries, although with widely differing results. It is evident that the economic structures of these countries vary greatly, which means that the opposing results depicted each generate a different balance of effects. There is a very noticeable reduction in resource extractions in Germany, which is a result of the high proportion of coal extractions, and the reductions imposed on them by the scenario's measures. The difference in resource extractions in the other countries can be explained by the rebound effect: the higher the effect on the gross domestic product, the lower the reductions in resource extraction.

Looking at the table we can see that a uniform program for Europe has very different economical and ecological effects on each country. The majority of electricity generated in France is atomic energy, yet in Germany coal is the most important source of energy, although there is a sharp increase, albeit still modest, in the use of renewable energy sources such as wind power. The differences in economic structures are also considerable, and have a decisive effect on the results: with its strong links in the global economy, a high proportion of Germany's added value lies in the manufacturing industry, whereas this sector is significantly smaller in Great Britain. The labor market in Great Britain is predominantly a competitive market, where supply and demand governs the wage rate. In contrast, in Germany and France we have a bilateral monopoly on the labor market, which means that employers' associations and trade unions negotiate wages. There are many more differences that can be linked to the degree in which a country would be affected by, and react to, environmental policy measures. Using the results from the model's calculations, the reverse conclusion would be the following: if we want to achieve the same results in individual European countries, then we will have to tailor the necessary environmental policy measures for each country.

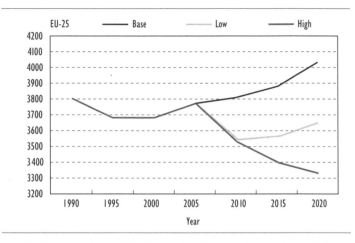

Figure 9   CO$_2$ emissions of the EU-25 for the "base" scenario, and the
alternative scenarios "high" and "low", in million of tons

The ecological effects for all of the EU-25 countries have been
summarized in Figs. 9 and 10. Figure 9 shows the advancement
in CO$_2$ emissions from the EU-25 starting in 1990 until 2020,
with the base projection and the alternative "low" and "high"
projections. If things continue the way they are currently pro-
ceeding, by the year 2020 CO$_2$ emissions will be 5.2% higher
than current levels, which would equal around 3.8 billion tons
of 1990's emissions, which are always the basis for international
target agreements. In the "high" scenario only 3.3 billion tons
of CO$_2$ will be emitted, which is admittedly 18.2% less than the
base projection, but only 13.2% below 1990's emissions. If we
were to include other climate gases, the result would be close to
the targets that have been set.

Figure 10 shows the total EU-25 resource extractions. Once
again the temporal developments of the three scenarios arc

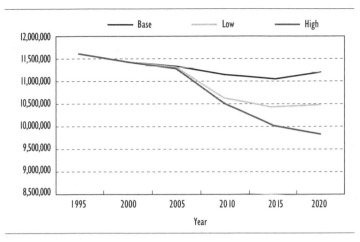

Figure 10 Resource extractions in the EU-25 for the "base" scenario, and the alternative scenarios "high" and "low", in thousands of tons

shown from the period of 1990 through to 2020. For the base projection the model calculated a stagnation of natural extractions in Europe. This does not mean that we have already reached a certain level of sustainability. We have merely come to realize over the past years that we can let extractions be made in the countries where we either directly import the raw materials, or indirect raw material imports contained in imported goods.

In the "high" scenario, by the year 2020 reductions will have barely reached 15%, although there would be a considerable increase in the gross domestic product resulting from this scenario.

We can conclude that an innovation strategy for Europe would deliver both economical as well as ecological advantages. However, when designing the measures, the different structures of each country will have to be taken into consideration. Therefore it would be better for each country to choose the instruments it wishes to combine with emissions trading, which has already been implemented.

The results from the calculations also show that the necessary changes will be anything but easy to implement. After all, in the "high" scenario, the price for a ton of CO_2 is 130 euros, which is quite high. If in the coming years we want to reach our targets using lower CO_2 prices, from between thirty to fifty euros, then we will have to introduce technical standards for household appliances, vehicles, and newly constructed buildings, using the variation of benchmarking discussed here. Moreover, the modernization of older buildings potentially offers substantial savings to CO_2 emissions. For example, we would have to additionally ramp up Germany's subsidization programs, which have been in place for some time now.

The global perspective

How would Europe's solitary steps towards a more sustainable development affect the world? In Figs. 11 and 12 we see the effects the "high" and "low" scenarios would have on global CO_2 emissions, and extractions of raw materials. Both illustrations contain the developments in quantities over a period of time in the base scenario, as well as the alternative "high" and "low" scenarios. The divergence of the alternative scenarios compared from the base scenario are portrayed on the right side of each illustration.

One look at Fig. 11 shows that measures that are restricted to Europe have a negligible effect on global developments. In the process, all of the developments in Europe's prices and income, which spread across the globe via international trade, have been included in the calculations. This is how the measures in Europe

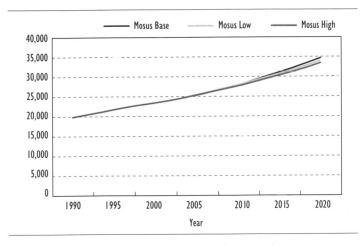

Figure 11 Global CO$_2$ emissions in millions of tons, in the scenarios "base", "high", "low", and the MOSUS project

bring about a reduction in Europe's demand for petroleum and thus for petroleum imports, for example, which then leads to a reduction of petroleum exports and production, as well as employment in Russia, OPEC, and other regions. Measures implemented in Europe also have a very indirect, far-reaching impact on the rest of the world. For example, increases in the price of European industrial products due to emissions trading, and the taxation of raw material extractions would hamper exports to the rest of the world and would have the tendency to increase imports to Europe, and consequently cause a rise in production as well as higher income in other global regions.

The divergence of the "high" scenario's CO$_2$ emissions from levels in the base scenario is hardly more than 3% in 2020. In the case of global raw material extractions (Fig. 12) it is also only 3.7%.

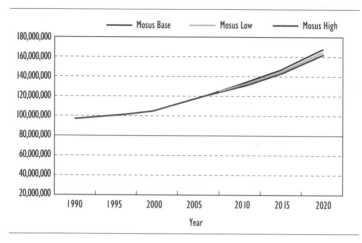

Figure 12 Global raw material extractions in the scenarios Base, High,
Low, and the MOSUS project in thousands of tons

The measures only apply to Europe, but naturally this is not
the case with regard to their effects. Table 12 shows the absolute divergence of raw material extractions, according to various
types of materials and CO_2 emissions in total global physical
units, and their distribution within Europe and over the rest of
the world.

Only 70% of the reduction of demands for primary energy
would take place in Europe, the remaining 30% would occur in
other regions of the world. In the latter case this mainly applies
to a reduction of imports of coal, gas and petroleum into Europe.
The global reduction of CO_2 emissions, totaling 961 million
tons, is divided into 72.4% in Europe, and 27.6% in the remaining regions. The reduction of CO_2 emissions outside Europe
has nothing to do with a decrease in fossil energy imports, since
CO_2 emissions only arise from the incineration of fossil fuels,

	World	Structure in percent	
		EU 25	Rest of the world
Primary energy (1000 tons oil equivalents)	−285,534	70.0%	30.0%
CO_2 emissions in 1000 tons	−961,000	72.4%	27.6%
Total raw material extractions in 1000 tons	−5,352,860	25.3%	74.7%
Biomass	−418,702	59.0%	41.0%
Coal	−2,265,356	34.7%	65.3%
Crude oil	−123,379	20.2%	79.8%
Natural gas	−60,232	37.9%	62.1%
Ore	−1,877,080	2.3%	97.7%
Other raw materials	−479,576	37.7%	62.3%

Table 12   The global effects of the High scenario of the MOSUS project in 2020, in absolute physical units, and their distribution across Europe and the rest of the world

Source: Giljum, St., Behrens, A., Hinterberger, F., Lutz, C., Meyer, B. (2008)

which obviously takes place in Europe. The cause for the reduction of CO_2 emissions outside Europe is the result of increased resource productivity within Europe, which leads to a decline in the amount of primary products imported. This results in a global reduction of the amount of goods produced, and subsequently CO_2 emissions.

Three-quarters of the reductions to raw material extractions, resulting from measures taken in Europe, actually occur outside Europe. This ratio highlights the extent of Europe's dependency on natural resources from the rest of the world. This is especially extreme in the case of mineral ores, where 98% of the reductions in extractions would take place outside Europe. Obviously this is due to the fact that hardly any ore is extracted in Europe today.

In this context we should also remember that in Germany, for example, the production of capital goods such as vehicles and machines constitutes the core of its economy – goods which are chiefly made of metals. But also in the case of fossil fuels, the effects would be higher in the rest of the world than in Europe.

We have demonstrated how these model calculations give a very detailed account of the economic interconnections between Europe and the rest of the world, divided into categories of goods and countries. Yet in the process, one factor has not been taken into consideration, a factor that is of great importance when estimating the significance of Europe's measures for the rest of the world. The proposed measures enable Europe to improve, among other things, the quality of vehicles, machines, and household appliances, which are, for example, built using lighter materials, would consume less fuel and could be utilized more efficiently. These capital goods and durable household appliances will, of course, also be exported to the rest of the world, in turn leading to increased raw material productivity in those countries that have not directly taken any measures to improve sustainability. It was not possible to include these effects in our calculations. With respect to vehicles and machines the impact would be considerable, since Europe holds a major share in this sector of the global market. Its share of the household appliance market is smaller, although it could also prove to be worthwhile for Asian suppliers to improve the quality of its products for its domestic market, and not only of its exports to Europe.

This does not render the results presented worthless, because the improved quality of Europe's export capital goods leads to a gradual increase in resource productivity outside of Europe. This is partly due to the fact that only a fraction of the capital goods used outside Europe are actually from Europe. Furthermore, it should be noted that the entire fleet of vehicles and machines in non-European countries would not be replaced all in one single step, but would rather undergo more of a gradual replacement of the older machines as and when the need for the use of newer machines arises. It is not yet possible to assess these difficult issues, although a new study using the GINFORS model is currently working on taking these effects into account. We can already conclude that exclusively European efforts, including the effects just mentioned, will not suffice to solve the global environmental problems.

9 Creating an International Framework

The alternative: no international framework

The damage mankind has inflicted on the environment is ultimately a global phenomenon. The image of spaceship Earth that we use to transport us through the universe, highlights the finiteness of resources and the limited options we have of storing pollutants. Improving sustainability in one country is not much help if, by doing so, problems in other countries increase. But this is what could happen if global courses of action are not coordinated. If one country imposes abatement costs on manufacturers, then sooner or later it is to be expected that there will be relocations to other countries that do not have such conditions, possibly leading to the much feared intensification of burdens on the global atmosphere.

An alternative could be to place the onus on consumers. One could impose a commodity tax on the goods that cause environmental burdens, or force emissions trading on consumers, or even set technical standards for consumer goods and buildings. Consumers do not relocate as often as industries do, so it is possible that such measures might work. On the other hand, the extent to which such a policy could be imposed is questionable. Any democratic government pursuing this kind of policy will be asked by the opposition why consumers should be subjected to this burden, when other countries do nothing of the sort, which would therefore hardly make much of a difference to the environment. Even if it were possible to set CO_2 emissions in Germany to zero, this would only have a marginal impact on global climate protection. Going it alone could only work if the majority of a country's voters are convinced that a more sustainable development in their country would improve the quality of their lives.

Admittedly we are a long way away from such a situation, which is the reason it is so incredibly difficult to introduce environmental policy measures. This is why it is so readily argued that there would be positive economic effects arising from following such a strategy. We only need to think of the debates on innovation strategy that are very popular in Europe, and which we have mentioned here. The emphasis is placed on technical progress in order to improve the quality of products and production processes. This tends to be a dynamic point of view, it is not only the prices on current markets that count in competition, but it is the presence on future markets that also develops economic sustainability. In order for this to happen, it is necessary for new production plants to be built for these future markets, and which will only be profitable once the resource prices have reached a specific level. So investors need certain prospects capable of sustaining themselves, which will not be the case without an internationally coordinated framework.

There is a possible danger that nothing happens without an international referendum, because each country is afraid of the economic risks. On the international level, as far as political decisions are concerned, we have a similar situation to that which we identified earlier as being the root cause of environmental problems, in other words, the decisions each individual makes concerning his or her use of the environment. It is even more commendable, at least as far as climate policy is concerned, that the EU has broken new ground and committed itself to the single handed task of cutting its CO_2 emissions by 20%, and is also prepared to go one step further by raising its reductions to 30% if other important countries are prepared to join in. An agreement had already been concluded in 1997, during the United Nations conference in Kyoto, but nowhere near all of the countries have signed this agreement. The protocol triggered much international debate, which has been followed up by negotiations, fuelled in part by the EU's initiative. What can we expect from it all, and where are the difficult points in the negotiations? We would like to tackle these questions in the following section, by first taking a closer look at the Kyoto Protocol, and then discussing the obstacles standing in the way of quick agreements being reached for a post-Kyoto commitment. Obviously international environmental policy cannot be restricted to climate problems, but we can also learn by using these experiences to develop a framework agreement for other areas.

The first attempt: the Kyoto Protocol

The United Nation framework agreement on climate change, passed in 1992 in New York, intended reducing the climate gas emissions of industrial countries by the year 2000 to match the levels of 1990. In 1995, during the first meeting of the contract parties in Berlin, it became clear that these commitments would not be enough. A draft was made for a new treaty, which was passed in consensus during the 1997 UN conference in Kyoto, as a protocol of general agreement on climate change. In this protocol the industrial countries committed themselves to reducing greenhouse gas emissions from 2008 to 2012 by a minimum of 5% of 1990's levels. The exact measures to be implemented were decided upon during the 2001 conference in Bonn, and approved in the same year in Marrakech.

Governmental representatives signed at the conference in Kyoto, but according to international law, a binding commitment requires the vote by the country's government – the act of ratification. Thus, the coming into effect of the agreement had to be tied to certain conditions and terms. Just imagine if one country had ratified, but for some reason or other, none of the rest had followed. Obviously the country concerned need not be bound to the contract. This is why the Kyoto Protocol first became valid ninety days after at least fifty-five countries had ratified, whereby the countries involved had to include industrial countries whose joint CO_2 emissions total at least 55% of the overall emissions from all of the industrial countries. In 2002, the European states ratified the protocol. During Clinton's presidency, America signed the protocol, but it has not been ratified during Bush's presidency. Australia and Croatia have also signed the protocol, but not yet ratified it. It was not until 2004, with Russia's ratification, that the protocol became legally binding. As of today, a total of 170 states have entered into the agreement.

In Annex A, the Kyoto Protocol defines climate gases, aside from the most important gas CO_2, as being methane (CH_4), nitrous oxide (N_2O), hydrofluorocarbons (HFCs), perfluorocarbons (PFCs), and sulfur hexafluoride (SF_6), which due to their climate gas status can be calculated into so-called CO_2 equivalents.

There are very different obligations for the individual countries, or groups of countries as the case may be in Annex B. Europe must make reductions by 8%, and Japan of 6%, while the Russian Federation and New Zealand are allowed to keep their emission levels of 1990. Within the EU agreements have been made to share the burden. So in order to reach reductions of 21%, Germany, for example, will have to make considerable efforts, because since the breakdown of the old industrial structures of East Germany after 1991, the most inefficient production technologies no longer exist. The UK's reduction target of 12.5% is, on the other hand, considerably lower. Spain has even been permitted to expand its emissions by 15%, because its level of industrialization is not yet fully developed, and consequently its original emissions level is still relatively low.

As far as the measures each country uses to reach their targets are concerned, the Kyoto protocol makes absolutely no stipulations. In fact, in Article 2 there is a catalog of measures itemized, which are to be viewed more as a listing, open to additions. Some of the measures included are: improvement of energy efficiency; reforestation; the promotion of more sustainable forms of agricultural cultivation; the study, promotion, and increased use of renewable energies, of technologies to absorb carbon dioxide, and of advanced environmentally compatible technologies; the implementation of market economy instruments and the elimination of subventions that lead to greenhouse gas emissions.

Article 6 regulates that two or more countries may run cooperative projects to reduce emissions, and then exchange a proportion of the subsequent results. This regulation, called Joint Implementation, has been designed in view of major differences, on the one hand, of economic power, and on the other between the costs of reducing greenhouse gas emissions in western industrial countries and east European industrial countries. Besides the usual measures to increase energy efficiency, these joint projects may also include setting up so-called "sinks." Sinks are facilities designed to absorb emissions. These mainly include reforestation measures. The additional trees extract CO_2 from the atmosphere and produce oxygen. The construction of nuclear power plants is a measure excluded from Joint Implementation.

Because the target agreements only cover commitments made by industrial countries, unless additional regulations are made, the potential reductions of the developing countries remain untapped. This is why Article 12 contains a so-called Clean Development Mechanism (CDM). When the investments an industrial country makes in a developing country demonstrably reduce that country's emissions, these reductions may then be charged to the industrial country's account. This process also has the added benefit that technologies are transferred from the industrial countries into developing countries. The facilities have to be certified by an independent third party accredited by the UN. The CDM measures may include the facilitation of sinks, but are exclusively restricted to reforestation measures, whereby no more than 1% of the industrial country's total emissions during 1990 may be made each year during the commitment period (2008 until 2012). The setting-up of nuclear power plants is excluded from the framework of CDM. The proceeds from the CDM measures will have to be used in part to cover the administration costs of the CDM. A further 2% of the proceeds are paid into a fund that will be used to finance the adaptation costs of climate protection in the particularly poorer countries.

Article 17 permits the countries committed to making reductions – the industrial countries – to trade with emissions allowances. If a country manages to keep its emissions below the agreed levels, it may sell its remaining allowances. Conversely, if a country is unable to fulfill its commitment, it may buy additional allowances.

In the case of targets not being met, Article 18 makes the contract parties responsible for enforcing the regulations. During the conference in Bonn, the following decisions were made: if after

the completion of the initial commitment period a country does
not have enough emissions allowances to meet target agreements
(including any and all additional purchases), sanctions will be
imposed. The shortfall from the first period will be multiplied
by the indemnification factor 1.3 and deducted from the country's emissions allowances for the follow-up period, and the
country will be excluded from emissions trading. Furthermore
the country will have to present a plan of action to the United
Nations, showing how future emissions reductions are expected
to be achieved.

The protocol has been heavily criticized, above all because
many consider the reduction targets to have been set too low.
This is problematic in two respects. If this criticism is meant in
the sense that the Kyoto Protocol will not be able to solve climate
problems, it is certainly correct, although it completely oversees
the dynamic character of the entire project. It is only the beginning of a long-term project, as can be seen with the target time
frame of 2008 to 2012. Obviously successive agreements will
have to set higher targets. In this way, we are taking smaller steps
to reach a bigger long-term goal.

If the criticism means that the targets set for the commitments
time frame are not ambitious enough for the countries participating, then it would help to take a look at the developments
that have taken place since 1990. Figure 13 shows the relative
divergence of climate gas emissions in CO_2 equivalents, from
1990 until 2004, of the Kyoto Protocol signatory countries. The
entire EU difference of 0.6% means that there is a long way to go
before target reduction levels of 8% are met. Bearing in mind the

Figure 13 The relative divergence of climate gas emissions in CO_2
equivalents from 1990 to 2004 of the Kyoto Protocol signatory
countries

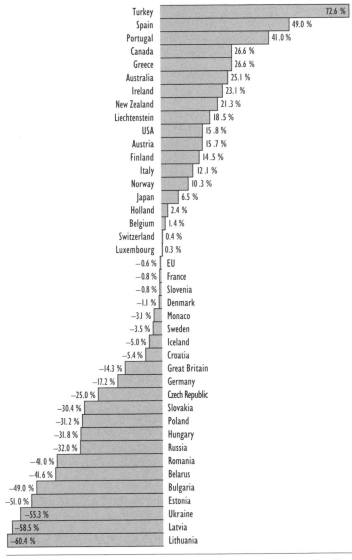

Turkey	72.6 %
Spain	49.0 %
Portugal	41.0 %
Canada	26.6 %
Greece	26.6 %
Australia	25.1 %
Ireland	23.1 %
New Zealand	21.3 %
Liechtenstein	18.5 %
USA	15.8 %
Austria	15.7 %
Finland	14.5 %
Italy	12.1 %
Norway	10.3 %
Japan	6.5 %
Holland	2.4 %
Belgium	1.4 %
Switzerland	0.4 %
Luxembourg	0.3 %
EU	−0.6 %
France	−0.8 %
Slovenia	−0.8 %
Denmark	−1.1 %
Monaco	−3.1 %
Sweden	−3.5 %
Iceland	−5.0 %
Croatia	−5.4 %
Great Britain	−14.3 %
Germany	−17.2 %
Czech Republic	−25.0 %
Slovakia	−30.4 %
Poland	−31.2 %
Hungary	−31.8 %
Russia	−32.0 %
Romania	−41.0 %
Belarus	−41.6 %
Bulgaria	−49.0 %
Estonia	−51.0 %
Ukraine	−55.3 %
Latvia	−58.5 %
Lithuania	−60.4 %

*Source: United Nations Framework Convention on Climate Change:
National Greenhouse Gas Inventory data for the period 1990–2004*

current revival of the European economic growth, it is certainly not going to be easy. For the individual European states, such as Spain, which have been granted an increase in emissions by 15%, the likelihood of these targets being adhered to is pretty slim, especially in view of the 49% increase already reached in 2004. On the other hand, the former socialist East European countries and Russia can boast sizable reductions that are way below the historic levels of 1990. It seems that these countries are profiting, as far as climate gas emissions are concerned, from the collapse of the old inefficient industrial structures. Figure 13 clearly shows how possible emissions trading might work between individual EU countries and other western industrial countries as demanders, and Russia and the former socialist East European countries as suppliers of emissions allowances.

The main problem is that the biggest emitter of climate gases – the USA – is not participating. This could be viewed as a failed climate convention. It could prove to be especially detrimental in view of persuading developing countries to accept emissions targets in the future. Consequently the Kyoto Protocol only applies to the European countries, Russia, New Zealand, and Japan. The developing countries had never been considered to be candidates for the commitments. Despite the Kyoto Protocol being restricted, more or less, to a European-Russian-Japanese affair, its importance should not be underestimated. The first steps have been taken along the path to long-term development, which could alone lead to restricting global warming by no more than two degrees. Emissions trading between the countries, Joint Implementation, and the Clean Development Mechanism, all these instruments still offer enormous potential. The integration of the United States and also the developing countries is crucial for the future. The European strategy of making concessions in order to step up negotiations could prove to be very helpful in this context.

The problem with balancing the interests of the developing countries, the newly industrialized countries, and the industrial countries

Right at the beginning, in Chapter 2, we concluded that by the second half of the century, we will have to stop continuously emitting CO_2 into the atmosphere if we want to prevent the increase in the Earth's average temperature from exceeding two degrees. This means that the total amount of global CO_2 emissions should not exceed the amount that plants can absorb via photosynthesis, and other processes. Natural scientists estimate that this would amount to emissions 80% below the levels of 1990, which would equal 4.1 billion tons. In order for this to happen, reductions no lower than 20% will have to be achieved by 2020, reaching beyond the Kyoto Protocol. Our final conclusion in Chapter 8 was that single-handed European efforts will not suffice to solve the climate problems. It will be decisive to involve both the United States, as well as the larger newly industrialized countries China and India, in climate agreements. According to the base projection of the MOSUS Project using the GINFORS model, by 2020 both China and India together will emit as much CO_2 as North America and Europe combined. If we were to add the remaining developing countries, then the industrial countries are noticeably in the minority. Because, however, the USA produces around half the emissions of the industrial countries group, it becomes obvious that a future climate regime only makes sense with American participation.

The developing countries argue against using units of emissions per capita, and do not believe they should need to take any measures, since in 2005 China emitted only 3.7 tons, and India only a mere 1 ton of CO_2 per capita, whereas the USA emitted 19.9 tons and the EU25 an average 8.3 tons. Seen from the fairness point of view, it is hard to disagree with this argument. The need for a unilateral obligation by industrial countries becomes even clearer if we consider that the volume of climate gases emitted into the atmosphere since the outset of industrialization originates almost entirely from the industrial countries. It would be unrealistic to think that the developing countries would overlook this point.

Seen from this perspective it becomes clear that incentives need to be given in order to convince the developing countries to join forces in reducing climate gases. In other words: it has to be worthwhile for the developing countries. The incentives could be highly varied. The simplest form imaginable would be to transfer payments from the industrial countries to the developing countries, on the condition that the beneficiaries agree to commit to emissions restrictions. It is imaginable that the core of future agreements with other industrial countries could be an extension of the emissions trading scheme practiced in Europe. A portion, at least, of the certificates should be auctioned off instead of being allocated out. A fund could be set up from the auction proceeds to finance the transfers for the developing countries.

Many discussants assume that the European ETS might be a nucleus for a world wide carbon system. But it seems to be unrealistic to expect a world system with a unique carbon price, as would be ideal from a theoretical point of view. For many countries such a cap would be too narrow and they would fear incalculable negative economic consequences. It seems to be more realistic that a hybrid system will emerge with isolated systems of pollution rights and tax regimes. Politicians are less interested in the ecological efficiency of an instrument, but they want to know the economic effects before the instrument is in action. So it can be assumed that in the international discussion there will be a bias for taxes.

The European Union has announced that Europe will follow

stronger CO_2 reduction targets, if the other countries cooperate. In this context the introduction of a world wide material tax which covers all material inputs including the fossil fuels should be addressed. The potential of a global material tax is higher than that of a global carbon market, because the material tax will reduce CO_2 emissions and will also help to solve many other environmental problems. Furthermore, especially for developing countries, the refunding of the revenue of the material tax is interesting: a reduction of taxes on consumption goods will have positive distributional effects.

A second possibility would be for the industrial countries to impose taxation on imports from those developing countries that do not comply with emission reduction targets. This would then bring about competitive pressure, which could increase the willingness to participate in the climate regime. However, this would place a considerable strain on the north-south conflict, and the developing countries might view the measures as the first steps in an economic war. It is also questionable how compatible the measures would be with the World Trade Organization's basic principles of free trade.

A third course could be to not expect the developing countries to formally commit themselves to emissions targets, but instead to expect them to be willing to cooperate with facilitation of technological transfers between the industrial and developing countries. The main philosophy behind this hypothesis is that the costs of reducing emissions in the developing countries are noticeably lower than in the industrial countries. The instruments could be those already introduced within the framework of the Kyoto Protocol, namely Joint Implementation (JI) and Clean Development Mechanism (CDM). A case of Joint Implementation could involve, for example, a German power supply company, subjected to German emissions trading, constructing a power station in China with a Chinese partner. Half of the difference in emissions saved from the substitution of the old power station would be credited to the German company, which would increase its stocks of emissions allowances, while the other half would go to the Chinese company, which could then sell its certificates on the market. A conceivable CDM set up might look like this: a German energy company makes some investments in China, without working with a Chinese partner. The company would be allocated the same amount of certificates that matches its emissions reductions in China, which it could then sell in Europe. The developing countries would benefit from the advantage of increased industrial efficiency, with no extra costs.

Research on the effect of the various policies on global pollutant emissions, and the economic development in the industrial and newly industrialized countries has yet to be completed. There is still a great need for information in order to be able to fully understand the various trade options, which will subsequently require researching. Each of the trade options mentioned is generally conceivable, along with several others, which we were unable to discuss here in detail. The option of the developing countries joining in the climate regime could be combined with the option of intensifying the transfer of technologies through JI and CDM. On the other hand, the option of taxing imports from countries not participating in the climate regime is problematic.

The small step taken by the Kyoto Protocol needs to be followed soon by a bigger step. Negotiations are already taking place at various levels, such as at the G8 summits in Heiligendamm and Tayako, and in the United Nations conferences at Bali and Copenhagen. Europe has taken on a leading role in this process. Now and in future, it will be essential to unwaveringly continue the course that has been chosen.

10 Final Comments

We should warmly welcome economic growth of the developing countries, because this is the only chance of eliminating the devastating poverty which is the ultimate cause of the third world's persistent population growth. However, the world will need different consumer goods and new technologies for a more efficient utilization of resources, unless we all want to expire as a result of an environmental catastrophe. Europe has to make a decision on a systematic innovation strategy, in order to globally initiate the processes of reducing resource utilization. In this book we have advocated a thesis which includes both sufficiency as well as efficiency strategies. When we say 'innovation strategy,' however, we do not mean that we can place all of our hopes on technological improvements alone. We also need to develop and demand new consumer goods that help save resources, which does not mean that we will have to reduce overall consumption.

The decisive question is how will we be able to actually achieve this? Now that the world is slowly realizing which destination we should aim for, the more complex question arises over the choice of instruments which need to be on our agenda. Quite a few things have already happened, but this is only the beginning. In this book I have proposed that economic instruments are undoubtedly important, but due to present flaws in the market, other instruments – such as information and communication instruments – as well as intelligent forms of regulatory policy need to be added. We have taken a detailed look in this book at these policy variants.

Due to existing economic structures, Europe is in a good position to follow these innovation strategies. But without an accompanying education campaign, in the long term, we will not have enough highly qualified employees at our disposal. On the other hand, many individuals will not be able to meet the increased demands of an ever increasingly complex working environment, despite further education campaigns. This is why we need an effective safeguard for the lower income groups, which could be achieved, for example, with the help of negative income tax.

Up until now, the climate debate has made astonishing progress in the discussions of environmental targets. But we should not fool ourselves; only when the necessary discussions over expanding the instruments of environmental policies necessary to attain our goals have taken place (which, as of December 2008, hasn't been the case), will we have any clarity as to whether we are on the right track. Moreover, we must also remain well aware of the fact that all our activities in Europe will be in vain if we do not succeed in implementing internationally binding agreements. Let us hope that policy makers will keep the momentum and achieve these critical objectives.

Glossary

Aachen Scenario: An information and communications policy scenario to reduce resource consumption. Its name originates from the Aachen Foundation Kathy Beys, which facilitated the first simulations of the scenario with the INFORGE and *PANTA RHEI* models.

Agenda 21: Agenda 21 is a developmental and environmental policy program geared to improving sustainable development in the 21st century. It was passed by 178 countries during the United Nation's 1992 Conference for Environment and Development in Rio de Janeiro.

Anthropocentric concept: A concept focused entirely on mankind.

Bionics: Bionics attempts to understand naturally occurring solutions to technical problems, and is concerned with applying these solutions to our technologies. Bionics is an interdisciplinary science, bringing together natural scientists, engineers, as well as architects and designers.

Biotechnology: The term biotechnology describes the incorporation of biological and biochemical knowledge to systems and products of a technical nature or application.

Blue Angel: Since 1978 the Blue Angel has been a seal of quality distinguishing particularly environmentally friendly products in Germany. In order to obtain the Blue Angel, an application must be made with the German Department of the Environment. An independent jury examines a product's suitability before the Federal Environment Ministry awards the seal of quality.

Business as usual: A conjecture of political behavior in simulation calculations with models. An assumption is made that policies will remain as they were in the past.

Capital goods: The part of a national economy's production that is added to the capital stock, instead of being consumed in the specific period.

CCS (Carbon Capture and Storage): A technical procedure used to separate CO_2 in coal-fired power stations, and its subsequent permanent underground storage.

Clean Development Mechanism (CDM): CDM is an agreement of the Kyoto Protocol, allowing companies in industrialized countries committed to reducing CO_2 emissions to invest in other projects which reduce CO_2 emissions in developing countries, and enabling the consequent reductions to be added favorably to their own target accounts.

Climate gases: Climate gases (also known as *greenhouse gases*) are responsible for the so-called greenhouse effect of the Earth's atmosphere: they absorb some of the infrared rays emitted from the Earth's soil, which would normally escape into outer space. The gases include carbon dioxide (CO_2), methane (CH_4), nitrous oxide (N_2O), hydrofluorocarbons (HFCs), perfluorocarbons (PFCs), and sulfur hexafluoride (SF6).

Computable General Equilibrium Models (CGE): A macroeconomic model, based on neo-classical economic theory, with a relatively detailed classification of markets and industries and numerically specified parameters.

Consumer goods: The portion of goods produced by an economy, during a specific period, and used by private households and public authorities.

Corporate governance: Corporate governance describes all international and national values and principles of responsible corporate management, and applies to both employees as well as management.

Cross section technology: See *Key technologies*

Disposable income: A household's income that remains after the redistribution with taxes and transfers.

Ecological tax reform (ecotax reform): Structural change in the tax system which raises taxation of natural resources and gives the revenue back by the reduction of other taxes or contributions.

Econometric models: Economic models with parameters based on historical data, obtained using statistical processes.

Economic instruments: Environmental policy sanctions offering monetary incentives to economic agents for more sustainable behavior.

E3ME: A multisector/multicountry economic-environmental model developed at Cambridge Econometrics (Cambridge, United Kingdom), with econometrically estimated parameters.

Emissions certificate: Pollution rights which can be traded nationally or internationally.

Environmental- Economic Accounting: An aggregate accounting system to depict the interrelations between the use of nature and economic development.

Eurozone: The region including all of the EU countries that have adopted the collective currency, the Euro. As of spring 2007, this included Austria, Belgium, Finland,

France, Germany, Greece, Ireland, Italy, Luxembourg, the Netherlands, Portugal, Slovenia, and Spain.

Fertility rate: The ratio between the average number of children and women of child-bearing age within a specific cohort.

Fossil fuels: Fossil fuels, such as coal, gas, and oil are the result of the decay of organic materials, originating from plant and animal matter millions of years old. Carbon dioxide gas is created when these fuels are combusted, and is then emitted into the atmosphere, causing the greenhouse effect.

Fuel cell: A fuel cell is an energy converter, which converts the chemical reactive power of a fuel and an oxidizer into electricity. Its technical application is mostly as hydrogen oxygen fuel cells in vehicles.

German Council for Sustainable Development: The Council for Sustainable Development supports the work of the State Secretary Committee for Sustainable Development in Germany. Its members are scientists, faith representatives, the business arena and the unions, as well as representatives from environmental and conservation organizations.

GINFORS (Global Interindustry Forecasting System): A global multisector/multicountry economic-environmental model developed at the Institute of Economic Structures Research (GWS, Osnabrück, Germany) with econometrically estimated parameters.

Grandfathering: The discretionary allocation of pollution rights based on past emissions.

Greenhouse gases: See Climate gases.

Gross domestic product (GDP): The total value of a region's finished products produced in a specific period, and measured in monetary units.

Gross national income: The income of a nation's citizens during a specific period in monetary units.

GTAP (Global Trade Analysis Project): A global neo-classical model developed at Purdue University with calibrated parameters, illustrating a detailed breakdown of economic and commodity group structures.

Human capital: An evaluation of a population's knowledge, ability and skills acquired through education, apprenticeship and training, and further education.

Industrial cluster: A concentration of specific industries within a region.

INFORUM International System: Globally linked system of economic country models with a sectoral disaggregation and econometrically estimated parameters led by INFORUM (University of Maryland, USA).

Infrastructure: A characterization of all long-term personnel, material and institutional facilities necessary for a modern division of labor in an economy. Occasionally it only applies to state provided services, such as a functioning judicial system, transport systems and public administration.

Inter Governmental Panel on Climate Change (IPCC): The IPCC was founded in 1988 by the United Nations Environment Programme (UNEP) and the World Meteorological Organization (WMO). The organization consists of hundreds of independent scientists from various branches, who team up to process research results on climate change which have been documented and published in scientific magazines, and issue a periodical report.

Intermediate products: That part of the production of an economy, which in the same period is used as input in the production process.

International Energy Agency (IEA): The IEA is an autonomous body within the OECD engaged in analyzing the energy market and investigating energy policy propositions. The

IEA was founded in 1974, and like the OECD, resides in Paris.

Intrinsic motivation: Intrinsic motivation is the term given when an individual's incentive for a specific behavior arises of its own accord, without the presence of monetary enticement or other external incentives.

Institutional investors: Institutional investors are establishments in which savings are turned into sizeable capital. The establishments include commercial banks, savings banks, building societies, investment companies, pension funds and insurance companies.

Joint Implementation (JI): JI is a facility designated by the *Kyoto Protocol* enabling companies in one industrial country to also invest in CO_2 reductions in another industrial country, which can be counted to meet the target in the first country.

Key technologies: Technological components used in many industries.

Kreditanstalt für Wiederaufbau (KfW): (Development Loan Corporation) is a German state owned bank. The KfW handles transactions for economic sanctions, such as promoting medium-sized companies and start-up businesses, granting small and medium-sized companies investment loans, as well as the funding of infrastructure projects and housebuilding, and the funding of energy-saving technologies and communal infrastructure.

Kyoto Protocol: The Kyoto Protocol is a supplemental protocol agreed upon during the 1997 United Nation's Framework Convention on Climate Change in Kyoto. This was the first time that binding targets were laid down concerning the greenhouse gas emissions of industrialized countries. The protocol came into force in 2005, and expires in 2012.

MOSUS: MOSUS (MOdeling SUStainability for Europe) is a research project which took place during the fifth EU research program, investigating the effects of measures to improve resource productivity in the European economy and its environment by using the simulation calculations of the global environmental economy model *GINFORS.* Twelve European research bodies from eight countries were involved in the project.

Nanotechnology: Nanotechnology involves the research and construction of very small structures: one nanometer equals one millionth of a millimeter. Nano (Greek *nanos*: dwarf). It is applied to energy technologies (fuel cells and solar cells), environmental technologies (material cycles and waste management), information technologies (new processors and storage), as well as in the health sector.

Natural capital: Natural capital is the metaphor used to describe stocks of natural resources. It does not necessarily have anything to do with the monetary value of these resources.

Negative income tax: Negative income tax is the term applied to a method of tax reform whereby income below a certain level, instead of being taxed, is topped up by supplements. Starting from a basic provision, the amount transferred decreases when earned income increases, so that the sum total of income and supplement grows with the increase of earned income.

Ordoliberalism: A market based economy system in which the state is obligated to ensure a proper legal environment for private property, freedom of contract, free competition, as well as maintaining monetary stability and social justice.

Organisation for Economic Cooperation and Development (OECD): The OECD brings together the governments of countries committed to democracy and the market economy from around the world to support sustainable economic growth, boost employment, raise living standards, maintain financial stability, assist other countries' economic development, and contribute to growth in world trade.

PANTA RHEI: A German economic- environmental model with econometrically estimated parameters characterized by a detailed classification of economic sectors.

Personal carbon trading: A concept introducing emissions trading to private households.

Post-Kyoto commitment: An international agreement on climate protection set to replace the Kyoto Protocol.

Process innovation: Facilitates new and more efficient production methods.

Regulatory policy instruments: Measures of economic and environmental policy that employ restrictive and prohibitive regulations.

Renewable energy: Inexhaustible energy obtained from naturally occurring energy processes. The energy sources are ultimately sunlight, the Earth's rotation, and geothermal heat. The energy is extracted and collected by wind power plants, solar power plants, tidal power plants, geothermal power and the use of biomass.

Renewable Energy Sources Act: A law promoting renewable energies. Plant owners are guaranteed a fixed reimbursement rate over a specific period of time to help secure the plant's profitability. The subsidy rate is reduced annually to provide a stimulus for more innovation.

Resource productivity: The relation between *gross domestic product* in constant prices and resource utilization measured in tons. The dimension used is currency unit per ton.

Rio Declaration: An environmental and developmental policy document, containing a preamble and twenty-seven principles, passed during the United Nations conference in 1992.

Social capital: A metaphor used to describe the total body of institutions and regulations which uphold social balance.

Social market economy: A market economy with governmental regulations geared to guaranteeing social balance.

Strong sustainability. A definition of sustainability according to which ecological and economic sustainability are not substitutable.

Sustainability: Sustainability is a normative concept which only permits developments that are able to meet current needs, without endangering the needs of future generations.

System of Economic Environmental Accounting (SEEA): An aggregate accounting system used by the United Nations to depict the interrelations between economic development and the use of nature.

Top runner: The norm set for a technical product, based on the best product available on the market. All market participants are officially required to reach this norm after a set period of time.

Weak sustainability: A definition of sustainability according to which ecological and economic sustainability are substitutable.

WTO: The World Trade Organization is an international organization based in Geneva, and is concerned with the regulation of economic and trade relations.

Bibliography

2: Where Is The World Heading?

Hahlbrock, K. (2007): *Kann unsere Erde die Menschen noch ernähren? Bevölkerungsexplosion – Umwelt – Gentechnik.* Fischer Taschenbuch Verlag. Frankfurt am Main.

Kemfert, C. (2007): "Breites Maßnahmepaket zum Klimaschutz kann Kosten der Emissionsminderung in Deutschland deutlich verringern." *DIW-Wochenbericht*, Nr. 18/2007. pp. 303–307.

International Energy Agency (2006): *World Energy Outlook.* Paris

Latif, M. (2007): *Bringen wir das Klima aus dem Takt? Hintergründe und Prognosen.* Fischer Taschenbuch Verlag. Frankfurt am Main.

Lutz, C., Meyer, B., Wolter, M. I. (2009): "The Global Multisector /Multicountry 3E-Model GINFORS. A Description of the Model and a Baseline Forecast for Global Energy Demand and CO_2-Emissions." In: *International Journal of Global Environmental Issues*. Will be published shortly.

Münz, R. /Reiterer, A. F. (2007): *Wie schnell wächst die Zahl der Menschen? Weltbevölkerung und weltweite Migration.* Fischer Taschenbuch Verlag. Frankfurt am Main.

Population Division of the Department of Economic and Social Affairs of the United Nations Secretariat (2005): *World Population Prospects: The 2004 Revision. Highlights.* New York.

Schellnhuber, H. J. (Hrsg.) (2006): *Avoiding Dangerous Climate Change.* Cambridge, Cambridge University Press.

Stern, N. (2007): *The Economics of Climate Change. The Stern Review.* Cambridge University Press.

3: What Are the Root Causes, And What Kind of Solutions Do We Have?

Bartmann, H. (1996): *Umweltökonomie – ökologische Ökonomie.* Kohlhammer. Stuttgart, Berlin, Köln.

Cansier, D. (1996): *Umweltökonomie, 2. Auflage.* UTB Taschenbuch. Stuttgart.

Ekins, P., Barker, T. (2001): "Carbon Taxes and Carbon Emissions Trading." In: *Journal of Economic Surveys*, Vol. 15(3), pp. 325–376.

Meyer, B., Bockermann, A., Ewerhart, G., Lutz, C. (1999): *Marktkonforme Umweltpolitik. Wirkungen auf Luftschadstoffemissionen, Wachstum und Struktur der Wirtschaft.* Physica-Verlag. Heidelberg.

Stern, N. (2007): *The Economics of Climate Change. The Stern Review.* Cambridge University Press.

4: The Sustainability Paradigm

Coenen, R., Grunwald, A. (Hrsg.) (2003): *Nachhaltigkeitsprobleme in Deutschland. Analyse und Lösungsstrategien.* Edition Sigma. Berlin.

Diefenbacher, H. (2001): *Gerechtigkeit und Nachhaltigkeit. Zum Verhältnis von Ethik und Ökonomie.* Wissenschaftliche Buchgesellschaft. Darmstadt.

Pearce, D. (2005): "Nachhaltige Entwicklung. Der heilige Gral oder unmögliches Unterfangen?" In: Fischer, E. P. und Wiegandt, K. (Hrsg.): *Die Zukunft der Erde. Was verträgt unser Planet noch?* Fischer Taschenbuch Verlag, Frankfurt am Main.

Spangenberg, H. (2005): *Die ökonomische Nachhaltigkeit der Wirtschaft. Theorien, Kriterien und Indikatoren.* Edition Sigma. Berlin.

Statistisches Bundesamt (Hrsg.) (2006): *Nachhaltige Entwicklung in Deutschland. Indikatorenbericht 2006.* Wiesbaden.

5: What Options Are There For Increasing Resource Productivity?

Aachener Stiftung Kathy Beys (Hrsg.) (2005): *Ressourcenproduktivität als Chance. Ein langfristiges Konjunkturprogramm für Deutschland.* Books on Demand. Norderstedt.

Distelkamp, M., Meyer, B., Wolter, M. I. (2005): "Der Einfluss der Endnachfrage und der Technologie auf die Ressourcenverbräuche in Deutschland." In: Aachener Stiftung Kathy Beys (Hrsg.): *Ressourcenproduktivität als Chance. Ein langfristiges Konjunkturprogramm für Deutschland.* Books on Demand. Norderstedt.

Fischer, H., Lichtblau, K., Meyer, B., Scheelhaase, J. (2004): "Wachstums- und Beschäftigungsimpulse rentabler Materialeinsparungen." In: *Wirtschaftsdienst*, 84 (4) pp. 247–254.

Grunwald, A., Coenen, R., Nitsch, J., Sydow, A., Wiedemann,
    P. (2001): *Forschungswerkstatt Nachhaltigkeit. Wege zur
    Diagnose und Therapie von Nachhaltigkeitsdefiziten.*
    Edition Sigma. Berlin.
Schmidt-Bleek, F. (2000): *Das MIPS-Konzept – Faktor 10.*
    Knaur Verlag. München.
Schmidt-Bleek, F. (2007): *Nutzen wir die Erde richtig? Die
    Leistungen der Natur und die Arbeit des Menschen.* Fischer
    Taschenbuch Verlag. Frankfurt am Main.

6: What Precisely Needs To Change To Enable Increased Resource Productivity In Germany And Europe?

Bach, S., Bork, C., Kohlhaas, M., Lutz, C., Meyer, B.,
    Praetorius, B. & Welsch, H. (2001): Die ökologische
    Steuerreform in Deutschland: Eine modellgestützte Analyse
    ihrer Wirkungen auf Wirtschaft und Umwelt. Physica Verlag.
    Heidelberg.
DeCanio, S. J. (1998): "The efficiency paradox: Bureaucratic
    and organizational barriers to profitable energy-savings
    investments." In: *Energy Policy* 26(5), pp. 441–454.
Grubb, M., Neuhoff, K. (2006): "Allocation and
    competitiveness in the EU emissions trading scheme: policy
    overview." In: *Energy Policy* 23(4), pp. 1–14.
International Energy Agency (Hrsg.) (2006): *Energy Policies of
    IEA Countries.* Paris.
Meyer, B., Distelkamp, M., Wolter, M. I. (2007): "Material
    Efficiency and Economic-Environmental
Sustainability. Results of Simulations for Germany with the
    Model PANTA RHEI." In: *Ecological Economics*, 63(1), pp.
    192–200.

Newall, R., Jaffe, A. B., Stavins, R. N. (1999): "The induced innovation hypothesis and energy saving technological change?" In: *The Quarterly Journal of Economics*, 114(3), pp. 941–975.

7: What Changes Will Have To Be Made To the Labor Market And To The Social Security System?

Homburg, S. (2003): "Arbeitslosigkeit und soziale Sicherung." In: *Vierteljahreshefte zur Wirtschaftsforschung*. 1, pp. 68–82.

Hüther, Michael (1990): *Integrierte Steuer-Transfer-Systeme für die Bundesrepublik Deutschland. Normative Konzeption und empirische Analyse*. Berlin.

Meyer, B., Wolter, M. I. (2007): "Demographische Entwicklung und wirtschaftlicher Strukturwandel – Auswirkungen auf die Qualifikationsstruktur auf dem Arbeitsmarkt." In: Statistisches Bundesamt (Hrsg.): *Neue Wege statistischer Berichterstattung – Mikround Makrodaten als Grundlage sozioökonomischer Modellierungen. Statistik und Wissenschaft*, Band 10. Wiesbaden.

8: Perspectives for More Sustainable Development In Europe

Giljum, St., Behrens, A., Hinterberger, F., Lutz, C., Meyer, B. (2008): "Modelling Scenarios towards a Sustainable Use of Natural Resources in Europe." In: *Environmental Science and Policy*. Vol. 11, pp. 204–216.

Hertel, T.W. (1997): *Global Trade Analysis. Modeling and Applications*. Cambridge: Cambridge University Press.

Burniaux, J. M., Truong, T. P. (2002): "GTAP-E: An energy-
    environmental version of the GTAP model." *GTAP technical
    paper* No. 16.
Uno, K. (Hrsg.) (2002): *Economy – Energy – Environment.
    Beyond the Kyoto Protocol.* Kluwer Academic Publishers.
    Dordrecht.

9: Creating an International Conceptual Framework

Agrawala, S. (Hrsg.) (2005): *Bridge over troubled waters:
    Linking climate change and development.* OECD. Paris.
Böhringer, C. (2002): "Climate politics from Kyoto to Bonn:
    from little to nothing?" In: *The Energy Journal.* 23 (2), pp.
    51–71.
Grubb, M. (1999): *The Kyoto Protocol: A guide and
    assessment.* London. Intergovernmental Panel on Climate
    Change (2000): *Methodological and technological issues in
    technology transfer: a special report of the IPCC working
    group III.* Cambridge University Press. Cambridge.

Picture References

All graphics: Peter Palm, Berlin (except figure 4, source:
Aachener Stiftung Kathy Beys [2005]: *Ressourcenproduktivität
als Chance – Ein langfristiges Konjunkturprogramm für
Deutschland.* p. 25).